THE CAPACITY FOR
WONDER

WILLIAM R. LOWRY

THE CAPACITY FOR
WONDER

Preserving National Parks

THE BROOKINGS INSTITUTION
WASHINGTON, D.C.

About Brookings

The Brookings Institution is a private nonprofit organization devoted to research, education, and publication on important issues of domestic and foreign policy. Its principal purpose is to bring kowledge to bear on current and emerging policy problems. The Institution was founded on December 8, 1927, to merge the activities of the Institute for Government Research, founded in 1916, the Institute of Economics, founded in 1922, and the Robert Brookings Graduate School of Economics, founded in 1924.

The Institution maintains a position of neutrality on issues of public policy. Interpretations or conclusions in Brookings publications should be understood to be solely those of the authors.

Copyright © 1994
THE BROOKINGS INSTITUTION
1775 Massachusetts Avenue, N.W., Washington, D.C. 20036

Library of Congress Cataloging-in-Publication data:

Lowry, William R.
   The capacity for wonder : preserving national parks / William R. Lowry.
      p.   cm.
   Includes bibliographical references and index.
   ISBN 0-8157-5298-9 (cl. : alk. paper).
   1. United States National Park Service—History.   2. Canadian Parks
Service—History.   3. National parks and reserves—Government policy—
United States—History.   4. National parks and reserves—Government
policy—Canada—History.   5. Nature conservation—Government policy—
United States—History.   6. Nature conservation—Government policy—
Canada—History.   I. Title.
   SB482.A4L68   1994   94-9122
   353.0086'32'09048—dc20   CIP

9 8 7 6 5 4 3 2 1

The paper used in this publication meets the minimum requirements of the American National Standard for Information Sciences—Permanence of paper for Printed Library Materials, ANSI Z39.48-1984.

Typeset in Bembo

Composition by Harlowe Typography, Inc.
Cottage City, Maryland

Printed by R. R. Donnelley and Sons Co.
Harrisonburg, Virginia

For the Opekas

Author's Preface

I had already been camping in Rocky Mountain National Park for days when I finally got around to doing the more formal part of my research. Even then, on the way to an interview, I stopped off at the backcountry office to pick up a map and an overnight permit. After all, the June day was pretty, and the mountains looked so inviting. The interview went well, and the assistant superintendent, Sheridan Steele, was pleasant and informative. But I felt restless sitting in his office, eager to get on the trail. Occasionally my mind wandered toward the snow-capped peaks and blue skies visible through the window behind Steele's head and, inevitably, my eyes followed. Maybe he noticed. Suddenly, his voice changed in tone and got lower as if he wanted me to concentrate to hear so that I would remember his next comment. "I hope you write this book," Steele said, "I hope you do a good job, and I hope a lot of people read it, because people need to know what's happening to the parks." An hour later, I was on the trail. As I hiked, I thought about Steele's words and realized how clearly he had expressed my own hopes. And as my senses became attuned to the sounds and smells of the forest, they were thus joined by another sense, a sense of urgency.

The national parks have always stretched my own capacity for wonder. From wide-eyed stares at Yellowstone's bears out the back windows of my dad's Chevy as a boy to sheer, exhausted exuberance on the summit of Mt. Whitney as a man, I have been impressed, inspired, and even intimidated by the national parks. I have been fortunate enough to vacation, work, live, love, camp, backpack, hike, swim, raft, kayak, and even howl at the moon a few times in the national parks of Canada and the United States. I have always loved being in the parks and just knowing that they are there.

The sense of urgency came later, as my understanding of politics increased. Modest as that understanding may be, I have learned enough to realize that of all the threats facing the parks, the most important is the behavior of government officials. How the parks are established, managed, funded, and ultimately preserved is determined by actions taken within the political systems of the two countries. If we, the public, are to protect these magnificent places, we need to recognize and challenge the political actions that affect the parks. It is that knowledge that motivates this book.

I learned much about the parks by going to them, experiencing them on their terms, and talking to the people who manage them. I use personal anecdotes of my trips throughout the work, not to be self-indulgent but to illustrate how political actions affect us, visitors to the parks. My comprehension of these issues would not have been possible without the assistance of employees, like Steele, of the U.S. and Canadian park services. I interviewed dozens of employees on my forays into the parks and into the offices of both systems between 1991 and 1993. From Rob Arnberger, who fed me popcorn in Big Bend on the Mexican border, to Prince Albert Park's Paul Galbraith, who talked my ear off one evening on the edge of Canada's great northern forest, I benefited and learned from every one of you. Thank you for your time and your candor. I still remember starting to leave Bill Paleck's office in Saguaro one day after an interview, only to stop when he said, "You got a few minutes? We should talk a while." I look forward to another talk sometime.

Many others deserve more than just the simple acknowledgments that follow. Most of my interviews with public officials are listed in the Bibliographical Essay. Some understandably asked to remain anonymous but were helpful nevertheless. Other observers of the parks, such as Michael Frome, Paul Pritchard, Al Runte, and Bill Waiser, graciously bequeathed some of their knowledge to me.

While traveling between parks, I enjoyed the hospitality of the Bersells, the Blacks, the Browns, the Brunis, Janet and Bill Caler, Bob and Colleen Delaney, Leslie Eliason, Randy and Patty Ferris, Charles Franklin and Liane Kosaki, Chris Gilbert, Barry and Emily Goldman, Judy Goldstein and Richard Purkey, Kathy Harrison and George Hoberg, An and Larry Heimann, the Harnishes, Keith Lee and Kathy Methot, Hans and Martha Michelman, Rick Olguin, Larry and Barbara Rothenberg, Kathy Teghtsoonian, and the Webbs. Several people joined me on various sojourns, but my father deserves special mention. He joined me in the Canadian Rockies for two weeks during which it rained every night. He never complained, and he never failed to get a fire started. In important ways, he and BJ and Tom have been with me all the way in this project. I also thank the Opekas for giving me a place to write this and so much more. They know why this book is dedicated to them, and that is enough.

People at Washington University, in particular my department chairmen, Bob Salisbury and John Sprague, as well as Deans Israel and Wilson, provided employment, leave time, and summer travel money. In between teasing me about "research" boondoggles, colleagues and students have been quite supportive. Several deserve particular mention: Barry Ames, Gayle Corrigan, Jim Davis, Lee Epstein, John Gilmour, Larry Grossback, Arnold Heidenheimer, John Kautsky, Jack Knight, Carol Mershon, Gary Miller, Andy Sobel, and Andy Whitford. Pauline Farmer, Maria Hunter, and Jan Rensing contributed valuable assistance, especially with the occasional snack.

Last but not least, I thank the people affiliated with the Brookings Institution for putting this book together. Nancy Davidson believed in this project from the start. She and three anonymous referees provided extremely constructive advice. The staff members who worked on the book, notably Jill Bernstein, Laura Kelly, Deborah Styles, and Norman Turpin, were impressively professional and thorough. Mary Baumann proofread the book, and Charlotte Shane prepared the index. The mistakes that remain are mine alone.

Ultimately, I stayed in Rocky Mountain National Park for several days after my interview with Steele and my backpack trip. I knew I needed to get back to St. Louis to meet a deadline on another project, but I did not want to leave. On the morning I just had to get going, I lingered around my campsite for much longer than I normally do. I built a nice fire, cooked breakfast, stared at Long's Peak in the background, and even wrote a letter. While I was packing, someone walked by and mentioned that they had seen some Rocky Mountain goats up on the north side of the park the evening before. As I drove east out of the campground, I approached a fork in the road. I weighed my options. The goats were considerably out of my way, but I had not yet seen any up close. On the other hand, I had to get on the road quickly or I would be driving some brutal hours. In the end the decision was easy. I took the long route. After all, I thought to myself, the goats may not be there forever.

Contents

Figures

For a transitory enchanted moment man must have held his breath in the presence of this continent, compelled into an aesthetic contemplation he neither understood nor desired, face to face for the last time in history with something commensurate to his capacity for wonder.

F. Scott Fitzgerald, *The Great Gatsby*

Chapter 1

Paved with Political Intentions

"Today, decisions are made for politics instead of for the resource."
—*Yosemite Superintendent Mike Finley*

Just five days before the 1992 U.S. presidential election, a small item, buried in all the accusations and polls, appeared in the newspaper. President George Bush, ending a twenty-two-year debate, had overruled environmental objections and approved the construction of mile-long jetties into the ocean off North Carolina's ecologically sensitive Outer Banks. The jetties will use land on the Cape Hatteras National Seashore under the jurisdiction of the National Park Service (NPS). Proponents argue that the jetties will improve seagoing passage for vessels of the seafood industry. Opponents counter, and the NPS warned, that construction of the jetties could cause enormous harm to fragile lands. The president's decision, however, did not wait for environmental impact statements that might settle these questions. Rather, because the issue involved a federal project in a state that Bush had to win to salvage his bid for reelection, the president seized a political opportunity. The chairman of the local Republican party had written a letter in July to the administration warning, "I cannot deliver the votes this fall unless some answers are provided in this matter."[1] Even the Department of the Interior's spokesman admitted, "If that [permit] helps voters decide that they ought to elect the administration that made it possible, that's great."[2] When I read the story, I remembered what Finley and so many other NPS employees had told me. Another decision affecting an NPS unit had been made for political expediency rather than for the benefit of the resource.

The national parks of North America exist in challenging times. These remarkable places were set aside to be kept in relatively natural condition, preserving lands and events that still inspire our "capacity for wonder." Today, however, the parks are threatened by increasingly difficult problems arising both inside and outside their borders. Most important, the park services of both the United States and Canada, responsible for

1

managing those lands and facing those threats, are currently changing. Changes in these political agencies will alter the way the lands are managed and significantly affect the future of our parks. Like Fitzgerald's enchanted travelers, we are currently holding our breath, hopeful that the last time in history will not soon be past.

In this work I study the changes in the park services of Canada and the United States between the late 1970s and the early 1990s. What caused them to change, in what ways did they change, and what do those changes mean for the national parks and other units under their jurisdiction? Specifically, how do those changes affect the abilities of these agencies to preserve these public places? The analysis relates to important issues for public agencies in general. Why do the political environments of public agencies change, and how do those changes affect their behavior? These are broad questions, and I address them with a theoretical argument based in political science and organization theory. The major focus of the book, however, is on an empirical analysis of the national park agencies.

A comparison of the two national park services is revealing because, while the two agencies are similar in many regards, they have changed their relative emphases in recent years. Why? An agency's ability to pursue a specific goal depends on political consensus on the goal, political support for the agency, and the commitment of agency employees to that goal. Over the past fifteen years the two park services have been affected by changes in consensus on goals and political support.

In the United States, conflicts over the goals of the NPS have become increasingly contentious, while support for the agency from other institutions of government has diminished. As a result politicians have been able to move many decisions away from field managers and increasingly focus agency behavior on serving their short-term, electoral needs. As in North Carolina, park lands are developed or paved, and not even for good intentions, as the old saying goes, but for political ones.

The Canadian Parks Service (CPS) has recently been changed by a growing consensus on specific agency goals and an increase in support for agency behavior from other government actors. The CPS has used its clarified mandate and relative autonomy to allocate greater responsibilities to field managers, who increasingly emphasize the long-term preservation of natural resources.

Pursuing the Goal of Preservation

The phenomenon to be explained in this study, what political scientists call the dependent variable, is the changing behavior of the U.S. and Canadian national park services. More specifically, I study the pursuit of the goal of preservation. Why has one park service changed to be able to pursue it more effectively, while the other finds it increasingly elusive?

The Goal

Most prominent of the common features of the two agencies is the mission statement. The mission of an agency is "a widely shared and approved understanding of the central tasks of the agency."[3] The mission statements of the two agencies are nearly identical. According to the 1916 Organic Act, the NPS is to "conserve the scenery and the natural and historic objects and the wild life therein and to provide for the enjoyment of the same in such manner and by such means as will leave them unimpaired for the enjoyment of future generations."[4] The CPS mission is quite similar: "The parks are hereby dedicated to the people of Canada for their benefit, education and enjoyment . . . and such parks shall be maintained and made use of as to leave them unimpaired for the enjoyment of future generations."[5] Both agencies are to facilitate use of the parks while still preserving them in "unimpaired" condition.

Such a mandate creates inevitable tensions and potential conflicts among seemingly incompatible goals. As the NPS admits in its management guidelines, "This language lies at the heart of national park system management philosophy and policy, although its interpretation has not been without controversy or differences over the years."[6] An extreme interpretation of potential use of parks in both countries could include all forms of recreation, commercial development, and even resource extraction. At the opposite extreme, leaving parks "unimpaired" could be interpreted as prohibiting any form of development and even locking out visitors. Between those two extremes, pursuing both goals—use and preservation—is the crucial, continuous challenge for the two agencies.

THE ESSENCE OF NATIONAL PARKS.   First, the possibility of preservation is essential to our conceptions of national parks. Most other public lands (including national forests and national rangelands) are subject to use, whether by private or public interests, but what makes national parkland different is the ideal of a timeless preserve. Indeed, in the United States, the NPS has used this differentiation to retain its territorial jurisdiction in the face of competition from other agencies.

This ideal has existed since the first national park, Yellowstone, was established in 1872. Legend has it that a group of explorers to the area in 1870 conceived of the idea of a natural preserve while sitting around a campfire along the Madison River. Recent evidence suggests a less romantic version, that the Northern Pacific Railroad encouraged recognition of the Yellowstone wilderness as a tourist attraction. Either way, the land was to be preserved in something close to its current state. In establishing the park, Congress called for "the preservation, from injury or spoliation, of all timber, mineral deposits, natural curiosities, or wonders . . . and their retention in their natural condition."[7] As historian Richard Sellars argues, "But whether the national park idea sprang from a campfire discussion or a corporate boardroom meeting, both versions reflected a common concern—to preserve Yellowstone's scenic majesty and to prevent private land claims from closing the area to the public."[8] Webster's Dictionary defines a national park as land to be "set aside and maintained." The protection of parkland from destruction, excessive development, or other impairment is essential to the idea of a national park.

THE MOST DIFFICULT TASK.   Second, the preservation goal is the most difficult task facing the agencies. Preserving the natural conditions of a park for the future depends on resisting the pressures of the present. Pressures have always existed to use national parks and historic sites for purposes other than preservation. Park units have been subject to nearly constant pressure from powerful interests seeking to extract resources and to develop the land. As the opening anecdote about the Outer Banks suggests, those interests have often found receptive ears in political authorities seeking opportunities for short-term advantage.

The rewards from preservation, however, are generally noticeable only in the long term and are not easily measured. Furthermore, the interests supporting the preservation goal are often overmatched in specific situations. Support for environmental ideals like preserving public lands in North America is widespread but not salient. In other words, mobilizing counterpressure to demands for use in specific parks is often difficult. The vast majority of potential supporters for preservation consists of tourists and others who "want access to the parks but whose interest in them is seldom expressed politically."[9] Political scientists have long argued that support for public goods (like parks) is usually overwhelmed by demands from private interests (pursuing such activities as development) precisely because advocacy of public goods is relatively unorganized and unfocused.

Certainly exceptions to this pattern exist. Preservation-oriented interest groups have become much stronger in recent years and have affected outcomes in some situations. In most cases, however, political influence shifts agency behavior away from long-term preservation goals and toward more immediate gratification. Those managing the parks recognize that pursuit of the preservation goal will often run counter to the short-term incentives of their political superiors. As NPS official Sheridan Steele told me, "Politicians have a two-year horizon whereas ours is much longer because people in the parks love the resource."

THE FOCUS OF EMPLOYEES.   Third, as the definition of mission suggests, mission statements can be misleading if the goals of the organization do not equate with the goals of its employees. This is hardly the situation for either park service. An anecdote illustrates the point. I had asked NPS ranger Jack Neckels a question. In responding, he began reciting the NPS mission. Having heard it from so many other employees so many times before, I could not help but smile. He noticed my smile and asked, "You wonder how we know it by heart? I've got it tattooed on my chest." Most employees of both the NPS and the CPS know the mission statements, as Neckels does, by heart. The statutory mission of each agency is almost a mantra for employees, repeated verbally to justify actions, cited in writing to explain decisions, recounted mentally to guide behavior, and perhaps even tattooed on a few chests. As an NPS interpretation ranger told me, "Work in the NPS is often a quasi-religious calling with the Organic Act as scripture."

For employees within both agencies, the mandated goal of preservation is key. Most of the career employees in the two agencies have made parks their careers because they enjoy wild places and the outdoors. On the one hand, pay and working conditions are often less than optimal. On the other hand, employees realize the other benefits of their jobs and joke about being "paid in sunsets." A 1980 survey of NPS personnel showed that 84 percent felt preservation was "the major purpose" of the agency, while only 9 percent emphasized use.[10] Surveys of CPS employees have elicited similar responses.

A WORTHY ENDEAVOR.   Finally, I concentrate on preservation because I consider it a worthy goal of the park service agencies. In deeming this goal a worthwhile agency objective, I contend that the parks are to be used, but used in ways that do not impair the natural conditions of protected areas. Preservation enables visitors to enjoy the parks through natural experiences in relatively undeveloped spaces. Interpreted quite literally, appropriate uses are those that do not detract from the "wilderness quality and ecological integrity" of parks.[11] This view, vividly described as "mountains without handrails" by one observer, has also been espoused by former employees, justified by legal scholars, validated in judicial rulings, corroborated by historians, and reaffirmed by statutory reauthorizations in both countries.[12]

Changing NPS Behavior

From the time of its formation in 1916, the NPS has built a reputation as an effective and professional public agency. The NPS earned that reputation by, in most instances, attempting to pursue its difficult mission of facilitating access while still protecting set-aside areas. The ability of the agency to preserve natural places inspired envy and emulation in park systems throughout the world. Indeed, when the NPS was created, the British ambassador called the preservation mandate "the best idea you ever had."[13] As late as 1978, impartial observers applauded the agency's efforts: "Surely many ills remain to be corrected, but recent developments in NPS policy, planning, and management suggest renewed affirmation of the Service's 1916 mandate."[14] Finally, the agency's ability to protect public lands inspired the public's trust. The NPS has historically been rated in highly or moderately favorable terms by nearly 80 percent of respondents in Roper polls.

During the 1980s and early 1990s, however, the NPS became increasingly less capable of pursuing the preservation goal. NPS units were often more likely to be used for some economic or political purpose than to be preserved in their natural condition. NPS observers began to take notice. By the early 1990s, the agency was being criticized for a seeming inability to handle crowds, excessive attention to development plans, inadequate protection of endangered species, inertia in the face of external threats, susceptibility to special-interest politics, failure to slow the deterioration of infrastructure, the sagging morale, and an inability to spend limited funds wisely. An introspective 1992 report by NPS employees and analysts participating in the Vail seventy-fifth anniversary symposium concluded: "There is a wide and discouraging gap between the service's potential and its current state, and the service has arrived at a crossroads in its history."[15]

The agency's own personnel are the harshest critics of ineffective pursuit of the preservation goal. I recently conducted interviews with more than thirty NPS field employees. These employees, ranging from maintenance rangers to superintendents, together represent over 500 years of experience in the agency. They were nearly unanimous in expressing serious concerns about the state of the NPS. Veterans such as Finley worried openly that preservation of resources was no longer the agency's highest priority. Many expressed fear that the NPS was losing sight of its own history and philosophy. Others said morale was at an all-time low. While all were still dedicated to their jobs and committed to service, most felt frustration at change that seemed beyond their control. As one said, if current trends continue, "the Park Service will no longer be able to achieve its mission."

Changing CPS Behavior

CPS behavior displays a completely different evolution. For decades, the Canadian national parks agency (it has undergone several name changes) sacrificed the goal of preservation to satisfy economic and political demands for development and recreation. Observers writing in the early 1970s criticized the agency for inattention to research needs, failure to educate the public, inadequate planning, ineffective programs, and excessive attention to recreation demands. A 1973 study concluded, "The branch [CPS] has perhaps been too willing to react to simple visitor demands and too reluctant to lead public opinion to develop a clear concept of the function and purpose of national parks in society."[16]

Canadian park management has changed in the past two decades. As early as 1978 analysts began to notice improvements in several areas of park administration. These modest alterations of the 1970s were greatly accelerated in the 1980s and early 1990s. The agency has recently introduced innovative efforts in research, fire control, resource inventories, education, public involvement, and comprehensive system planning. The CPS has, in short, shown a renewed commitment to the statutory goal of preserving natural areas. As one recent study concludes about the CPS, the agency is performing quite effectively.[17]

The comments of agency employees corroborate this perspective. I recently finished a series of interviews with Canadian park officials at all levels of the organization. While many admitted that the agency had often ignored the statutory goal of preservation in the past, all were positive and upbeat about the agency's renewed focus. All expressed optimism about the reaffirmation of preservation as the agency's purpose. One summary comment was, "We aren't there yet but we are headed in the right direction."

A Recent Divergence

Over the past fifteen years, the behavior of the two agencies has diverged. I use a fifteen-year framework because that period roughly covers important events in each agency: placement of the CPS in the Department of the Environment and the beginning of the Reagan revolution in the United States. The NPS, in spite of the preferences of its own employees, has seemingly lost its capability to pursue preservation. The CPS increasingly pursues the same goal by managing parks to preserve them in as natural a condition as possible. Why has this divergence occurred?

Explaining the Changing Behavior of Public Agencies

An agency's ability to pursue a goal depends on the political consensus behind the goal, political support for the agency, and the commitment of its employees to mandated goals.

My explanation for the changes in the behavior of the two agencies is based on the assumption that agency employees are and remain substantially committed to the goal in question. This assumption is well justified in the cases of the park services and their commitment to the preservation goal and it is necessary, since employee commitment is crucial to efficacious agency behavior. As Kaufman described in his classic study, the Forest Service depended on employees who possessed not only the capacity but also the will to do their jobs.[18] Numerous other analyses of bureaucratic behavior have also recognized how essential employee commitment is to agency performance. High levels of commitment are not, however, enough to guarantee effective agency pursuit of a goal.

Goal consensus and political support are generally more susceptible to rapid change than employee commitment and thus provide the dynamic element behind shifts in agency behavior. Changes in goal consensus and support lead to alterations in the organizational structures and operating conditions of the agencies. Ultimately those changes modify the behavior of the agencies. If two agencies, even agencies like the park services, with identical missions and with equally committed personnel, are subject to political variables that change in different directions, then the two agencies will show increasingly divergent patterns of behavior.

Changes in the Political Environment

Two political factors are crucial determinants of agency behavior in that they are essential for effective pursuit of statutory goals. One has to do with the means of agency performance, an adequate level of political support for the agency. The second involves the ends of performance, the degree of external consensus on specific goals to be pursued. The levels of these two factors tend to vary together, but both consensus on goals and political support are necessary for effective, goal-oriented agency behavior. If the levels of these factors change, so too will agency behavior.

POLITICAL SUPPORT. Political support expresses itself in the supply of material resources, the relative autonomy given to an agency, public trust in agency motives, and the degree of institutional deference to bureaucratic expertise. Some observers note that external political support, particularly the support of Congress and the president, is essential to successful bureaucratic implementation of programs. Others contend, however, that the desire for autonomy may prevail over the need for material resources. "Organizations are often prepared to accept less money with greater control than more money with less control."[19] Agency attempts to maximize political support and minimize external threats are "the bureaucrat's paramount concern."[20]

This variable can change dramatically. Levels of political support for particular agencies change with administrations, events, personalities, and even conflicts over goals. "Political support is at its highest when the agency's goals are popular, its tasks simple, its rivals nonexistent, and the constraints minimal."[21] Political support is at its lowest when the agency has no insulation from external actors. The variation in political support is evident in such factors as level of budgetary resources, isolation resulting from structural placement in government, degree of interagency conflicts, and job security for agency leadership.

CONSENSUS ON POLICY GOALS.   Many classic studies have described the importance of clear, consensual goals. "Public agencies need a statutory charter that explicitly declares their primary policy objectives and estab-lishes priorities among them."[22] Even the seemingly simplest public projects will be more difficult if they are initiated with contradictory legislative criteria. Many of the most damning critiques of agency behavior begin by noting the existence of vague policy goals on which no meaningful con-sensus has been reached. Delegation of policy responsibility to agencies is inevitable, but "the practice becomes pathological, and criticizable, at the point where it comes to be considered a good thing in itself, flowing to administrators without guides."[23] The effects of vague, nonconsensual goals are important enough to have spawned a literature on why legislators often delegate without seeking agreement on explicit guidelines.

Even explicit mission statements do not guarantee consistent emphasis on specific goals. When consensus is not reached on the values behind goals or on the relative prioritization of potentially competing goals, the agency's clarity of direction is diminished. Subsequent agency determi-nation of priorities is then subject to constant criticism, further obscuring original intentions. Considerable variation exists in the degree of consensus behind the policy guidelines assigned to bureaucratic agencies. Kaufman's study of the U.S. Forest Service described an agency fully dependent on remote managers but able to pursue its mission by maintaining explicit, agreed-upon rules and policy guidelines. However, the opposite is also quite possible. Many have criticized the lack of consensus on the goals of the Office of Surface Mining Reclamation and Enforcement. "It is hard to find a good word for the design of OSM's statutory mandate."[24] The variation in consensus on goals for different agencies is apparent in conflicts over mandated tasks and confusion over agency intentions.

Changes in the Operating Environment

While these two political factors, support and consensus, are theoret-ically independent, changes in them often occur in reinforcing ways that are reflected in the agency's operating environment. "The skills of a par-ticular bureau are inextricably intertwined with its mission."[25] Political support is likely to be higher when consensus on goals is greater. Ambiguity or conflict in goals is conducive to ambiguity or conflict in political support.

When agencies enjoy consensual goals and high levels of political support, agency operations reflect independent decisionmaking, empowerment of relevant actors within the agency, and high employee morale. Bureaucrats are still accountable to the political process because they must attain their goals in order to retain their relative autonomy. They are, however, relatively free from political meddling in their agencies' affairs and can structure their agencies to maximize the pursuit of explicit goals without the need to be constantly looking over their shoulders. Agencies dealing with distinctive circumstances can effectively decentralize decisionmaking to those who understand the specific situations. Day-to-day decisionmaking authority can be delegated to the level at which knowledge and expertise in local circumstances and specific situations are highest.[26] Employees can be empowered to implement designated tasks. In his study of street-level bureaucrats, Lipsky noted the effect on morale of empowerment with clear responsibilities: "If goals were clearer, workers could direct their energies with less ambivalence."[27] Kelman also concluded that employee morale and effectiveness are enhanced by such operating conditions: "Perhaps the most important thing that could be done to improve the operating performance of government agencies is to increase the freedom and the individual responsibility of people working in them."[28]

Agencies beset by diminished goal consensus and the lack of political support are doomed to vastly different experiences. Lacking insularity and forced to choose constantly between different goals, bureaucrats will be subject to continual second-guessing and political intervention. Inevitably, disputes will arise both inside and outside the agency. Inside, the morale will suffer if employees are subjected to conflicting demands and frequent criticism. Outside, agency decisions will stimulate controversies with other concerned parties and agencies that may well have expected some other behavior based on past attempts to satisfy different goals. As more and more decisions are questioned, pressure builds within the agency to remove discretion from lower levels and to centralize authority and resources at levels that interact with external political actors. Day-to-day decisionmaking authority is subject to centralizing pressures based on higher-level fears of lower-level mistakes. The agency thus becomes even more susceptible to external intervention and less likely to develop its own organizational capabilities.

Changes in Policymaking

The changes described above in the structures and operating environments of public agencies will lead to changes in policy outcomes. "Structural choices have important consequences for the content and direction of policy, and political actors know it."[29] The following hypothesized effects of the operating environment on policy outcomes are based on two simple assumptions. First, the behavior of an agency is determined largely by who in the agency has decisionmaking authority. Other studies have demonstrated the influence of certain groups within an agency when they have relatively greater power than other groups. Katzmann's study of the Federal Trade Commission, for example, showed that the kinds of cases pursued by the agency depended on whether decisions were made by the lawyers in the Competition bureau or by the economists in the Economics bureau.[30] Second, one useful way to differentiate among decisionmaking groups in agencies is among those who affect the agency as political actors, most obviously appointees, and those who work within the agency permanently as careerists. These categories are important because the two groups respond to different incentives and concerns. Political authorities are likely to concentrate on short-term electoral incentives or ideological emphases, whereas careerists generally stress long-term agency goals.

The changes in the operating environment of the agency with consensual goals and high political support will shift decisionmaking authority to allow careerists to use long-term strategies to pursue goals that fit the agency's mission. When authority can be allocated to levels of the organization at which pertinent information is available, agency careerists can use their specialized expertise. Empowered employees will base their decisions on agency-defined concerns and will act enthusiastically. Behavior can be innovative and experimental. Accountability is determined by the long-range attainment of ideals explicit in the mission statement.

Changes in the operating environment of the agency with competing goals and little political support will shift authority to appointees and to external political actors. Agency behavior will be focused on satisfying political authorities in the short term. The increased intervention by political appointees and authorities will lead to more decisions based on political calculations. Innovative actions will be considered risky and therefore unlikely. Accountability will be determined on a case-by-case basis. In short, the agency will be forced to neglect pursuit of certain long-range statutory goals to serve other interests.

Explaining the Changing Behavior of the Park Services

How does this theoretical argument apply in the context of the two national park service agencies? The park services of Canada and the United States are quite comparable in important ways, not the least of which is the commitment of agency employees to the goal of preservation. In the past fifteen years, however, the CPS has become increasingly able to pursue this goal, whereas the NPS has not. This divergence results from recent changes in political support and goal consensus for the agencies. These developments stimulated changes in the operating environments and behavior of the CPS and NPS that are consistent with the theoretical argument discussed above.

Differences in Political Support

Institutional differences between the United States and Canada in political support for bureaucratic agencies have been exaggerated since the late 1970s.

INSTITUTIONAL DISSIMILARITIES.   Institutional dissimilarities between Canada and the United States involve fundamental aspects of the two political systems. Differences involve the style of government, the role of political actors, and the nature of federalism.

Canada and the United States use different formal styles of government. Canada uses a Westminster-style parliamentary system with single-member districts, disciplined political parties, and elected members of the parliament's majority serving as prime minister and cabinet ministers. The U.S. presidential system uses single-member districts as well, but the president and legislators are elected separately, and party discipline is often relatively low. Parliamentary democracies are, in general, more likely to foster cooperative relations between the bureaucracy and other political actors than are presidential systems.

Several factors are conducive to a greater level of political support for Canada's bureaucracy. Power can be heavily centralized within the prime minister and the cabinet, thereby creating the potential for majority parties to significantly control the bureaucracy, but this potential does not necessarily translate to political manipulation of agencies. A Canadian prime minister only uses about a hundred political appointees, whereas U.S. presidents use several thousand. Members of the majority party are often reluctant to oppose the actions of an agency headed by a fellow party member. Minority party legislators may question agency behavior, but, as opposition party members admitted to me, effective intervention is difficult. Entrepreneurial actions by individual legislators are made impractical by the tremendous difference in power between cabinet ministers and backbenchers in the Parliament. Indeed, surveys reveal that legislators are less likely to intervene in agency affairs in Canada than in the United States.[31] Overall, this Westminster framework enhances institutional deference to the bureaucracy, thereby nurturing relatively autonomous agency behavior.

The U.S. bureaucracy is more susceptible to political control. With U.S. separation of powers, public agencies can be pulled one way by the White House and another by the Congress. The bureaucracy is often caught between the branches and despised by both as "the institutional participant in the political process that everybody loves to hate."[32] Further, individual legislators in the fragmented, specialized U.S. Congress are more likely to intervene in agency behavior than are their counterparts in Canada. Historically, legislators have often attempted to micromanage agency affairs. Because individual members of Congress do have influence, political appointees must respond to their demands. Finally, political appointees in the United States are willing to use their agencies for the president's benefit since they have no electoral mandate of their own. Overall, political scientists characterize the U.S. bureaucracy as "relatively weak" and often "heavily penetrated."[33]

Other actors external to the bureaucracy, besides the chief executive and the legislature, can influence levels of political support for agencies either directly or through elected officials. Important differences also exist here between the Canadian and the U.S. systems.

First, public agencies in Canada have historically enjoyed higher levels of public respect than those in the United States. The Canadian bureaucracy has thus been less susceptible to public ridicule and media criticism.[34]

Second, Canadian interest groups have historically been less prominent than their U.S. counterparts in agency deliberations. In part their lesser role results from groups' often having been represented in political bargaining situations by members of the government, usually in the person of the head of the relevant agency. More generally, Canadian interest groups have been involved in agency decisionmaking situations, but their influence has been diminished by their lack of power in Parliament. As other analysts have noted, "interest-group pluralism has never reached the sanctity of belief and practice [in Canada] that it has in the United States."[35]

Third, Canadian agencies have been less susceptible to judicial intervention than in the more litigious United States. In general, the parliamentary system creates few institutional incentives for the courts to attempt to restrict agency discretion.[36] While Canadian courts have traditionally deferred to parliamentary authority, U.S. courts have for centuries had the potential, even if they have rarely used it, to intervene in agency actions.

A final institutional difference between the two countries involves their systems of federalism. In general, Canadian provincial governments are considered more powerful, more independent, and more influential than their U.S. counterparts. This difference is particularly noticeable in the policy area of public lands. The U.S. federal government enjoys much greater constitutional jurisdiction over lands than do central Canadian authorities. The U.S. government owns roughly one-third of the total land area in the country including forests, rangelands, and parks. The majority of public land in Canada has belonged to the provincial governments since the Constitution Act of 1867. Land for parks must be transferred from the provincial to the federal government. As one author notes, "Canada is unusual as a federation [in] the degree of power of the provincial governments over natural resource policy."[37]

RECENT EXAGGERATION OF INSTITUTIONAL DIFFERENCES.   If relevant institutional differences between the two countries had remained static over the past fifteen years, they would be less helpful in explaining the recent divergence between the CPS and the NPS. They have, however, been exaggerated in recent years, especially in the area of national park management.

Political support of Canadian agencies from elected officials has increased in the past fifteen years, particularly in terms of increased delegation and deference toward the bureaucracy. Despite Prime Minister Pierre Trudeau's aggressive attempts to centralize many decisions, trends toward delegating responsibilities to the public agencies actually accelerated during his last term (1980–84). These trends gained momentum in subsequent years. When Brian Mulroney's Conservative party took power in 1984, its themes included explicit efforts to increase managerial delegation and agency discretion. Those efforts climaxed with the release of *Public Service 2000*, or PS 2000, in 1989. PS 2000 mandates accountability, reduced upper-level controls, and greater trust in employees, or, as the official announcement declared, "the tools and the authority to perform effectively."[38]Thus, even though Canadian budgets in the 1980s did not increase material support for many agencies, the bureaucracy has received greater discretion and control.

During the same period, the political support given U.S. agencies by elected officials has diminished. Political appointees have greatly increased their control over public agencies. The Reagan administration (1981–89) centralized and politicized the bureaucracy more than any other modern presidency. The administration's efforts benefited from many factors, including the 1978 creation of the Senior Executive Service, but Reagan's actions actually constituted an acceleration of trends that had begun decades before.

The recent increase in political control is evident in the fact that the debate over Reagan's "administrative presidency" centers on benefits and costs rather than on whether the increase actually occurred.[39] Many remaining ideals of "neutral competence" in the bureaucracy were replaced with ideals of partisan loyalty.

U.S. civil servants were under pressure not only from the Reagan administration, however. With divided government (different parties controlling different branches) and intense ideological differences, Democrats in Congress responded to Republican presidential assertions with demands on the bureaucracy of their own. Legislators from both parties took advantage of opportunities to become involved in agency behavior. In an upward-spiraling cycle of institutional competition, political authorities from both the executive and the legislative branches intervened aggressively in the actions of public agencies. During the same period, budget cuts and concerns about the deficit prevented material resources for U.S. agencies from keeping pace with increased demands.

Differences in political support have also been exaggerated by other recent developments. First, the media and the general public remain more supportive of government actions in Canada than in the United States. Vietnam and Watergate contributed to a greater distrust of government in the United States than exists in Canada.[40] Canada "remains more respectful of authority, more willing to use the state."[41]

Second, though Canadian interest groups have become more active, especially since the 1982 Charter of Rights, groups remain less likely to intervene in agency affairs in Canada than in the United States. The most obvious reason for differences between interest groups in the two countries involves sheer numbers. Comparison of the five major environmental groups in the two countries, for instance, shows that those in the United States have nine times as many members as those in Canada and more than ten times greater financial resources. While some perspective must be maintained on the basis of population differences, the fact is that U.S. interest groups remain more noticeable and more influential than their Canadian counterparts. Furthermore, divided government facilitates access by interest groups to policymakers. Since the United States experienced divided government continuously between 1981 and 1993, the relatively greater degree of access for U.S. groups may well have been exaggerated.

Third, Canadian courts have also become more aggressive recently, but their intervention in agency affairs has not yet approached the level of judicial activism seen in the United States. The role of the Canadian courts was strengthened by the entrenchment of individual rights in the Constitution by the 1982 Charter of Rights. The charter formally guaranteed, in the constitution, the basic freedoms such as religion and association as well as rights against search and seizure, cruel and unusual punishment, and others. The courts have become increasingly involved in environmental cases, particularly those concerning dams, just since 1989. Still, the Charter of Rights contains huge loopholes and exemptions from judicial intervention. For example, the rights extend only to "reasonable limits." Furthermore, Parliament and provincial governments retained the power to contravene certain rights.

The United States in the past two decades has experienced an intensely "heightened judicial scrutiny of administrative action" through lowered standing requirements, increased burden of proof on agencies, and diminished deference to bureaucratic expertise.[42]

Fourth, while various factors have caused different levels of political support for each country's bureaucracy in general, differences in the federal systems, particularly in constitutional jurisdiction over land, have specifically affected political support for the parks. In the United States the diversity of federal land holdings resulted in the NPS being placed in and remaining in the multifaceted Department of the Interior. In Canada the situation was quite different, facilitating placement of the Canadian parks agency in a much more focused department. Transfer of the agency from the rather eclectic Department of Indian Affairs and Northern Development into the supportive Department of the Environment in 1979 increased support for and deference to the agency from other political actors.

Differences in Consensus on Agency Goals

The relative presence of explicit, consensual goals will vary much less systematically between Canadian and U.S. agencies than differences in political support. Traditionally, even while Canadian agencies enjoyed considerable discretion in implementation, they were generally assigned relatively specific goals. Political consensus on those goals was likely given the parliamentary framework and strict party discipline of the Canadian system. Overall, Canadian political authorities have specified policy objectives and then granted broad discretionary powers to bureaucrats for planning and strategies.

U.S. policymakers attained a certain notoriety for delegating with vague goals. As Lowi concluded about the post–New Deal United States, "it is no longer possible to require that broad delegations of power be accompanied by clear legislative standards."[43] In recent years Canadian delegations of authority have come to approximate more closely the traditional U.S. system in many policy areas. Scholars have noted "a weakening of the capacity and will of political actors to set goals for bureaucratic actors."[44] Reasons include the personalities or, some argue, the ineptitude of political leaders. This convergence between the two systems in goal obfuscation did not occur with the national parks agencies, however.

INCREASED CONSENSUS ON GOALS FOR THE CPS.   The trend toward obscuring agency goals did not occur in policies concerning Canadian national parks. In fact during the 1980s Canadian policymakers coalesced behind the agency goal of preservation of natural areas.

Why did consensus building center on preservation? My answer involves the dramatic increase of environmental awareness throughout Canada, the pragmatism of the Mulroney government, and the power of provincial authorities. Concern for environmental issues became a political imperative in the past two decades. Mulroney and other political leaders were well aware of the benefits of responding to these concerns. Responding by encouraging preservation of federal lands was much easier than addressing issues involving provincial lands and developments. Provincial lands host extensive logging, grazing, and recreational activities. Further, since provincial governments are considerably stronger than U.S. state governments, they are a force to avoid antagonizing politically. The power of provincial governments became particularly noticeable in the 1970s and 1980s when federal-provincial relations became increasingly contentious in many policy issues, including those affecting natural resources. In addition, the powers of regional ministers and provincial authorities increased dramatically during the 1980s. Finally, Mulroney needed the support of provincial leaders for his constitutional reform efforts. Rather than confronting territorial demands for continued economic development of provincial lands, the Mulroney government responded to environmental concerns about public lands by rhetorically emphasizing, though not consistently enforcing, the preservation of federal parklands in natural conditions.

DIMINISHED GOAL CONSENSUS FOR THE NPS.   In the United States the traditional lack of clarity in agency goals has intensified in the NPS during the past fifteen years. The potential for tensions within the NPS mandate was exaggerated by conflicting directives to the agency in the 1980s.

The uncertainty about goals in the NPS has its roots in divided government, in the power of organized interests in the United States, and in the increased awareness of threats facing the national parks. The Reagan and Bush administrations were hostile to many environmental concerns, and they were supported by western interests calling for the development of public lands. These administrations faced an opposition-controlled House of Representatives and (after 1986) Senate that supported ecological causes and were under pressure from an increasing number of environmental interest groups. Disputes were stimulated and magnified by serious threats facing the parks in the 1980s. During the decade the traditional balancing act between goals of preservation and use degenerated into open conflicts, and no consensus was developed behind a specific goal.

Hypothetical Results

Theoretically these changes in political factors will lead to changes in the operating environments for the two park services, which in turn will result in divergent policy behaviors. Changing operations shift the allocation of authority. Ultimately different sets of actors making decisions will have different emphases.

High political support and increasing consensus behind the goal of preservation in the CPS will foster independent decisionmaking and a high degree of employee enthusiasm and will discourage interagency conflict. The CPS will be able to plan systematically and to allocate authority to the level of the organization that has the relevant expertise. Many day-to-day operations will be decentralized to empower field-level employees. These bureaucrats, already partial to ecological concerns, will stimulate and expand rhetorical federal directives encouraging preservation, thereby reversing a decades-old policy of "paving" the parks through construction and development. Canadian parks will thus increasingly reflect a long-term focus on the agency's mission of allowing park use only in ways that preserve natural conditions.

The NPS will operate under conflicting goals bereft of political support. Political intervention and interagency conflicts will increase. Employee morale and trust in leadership will subsequently diminish. As involvement of political authorities increases, many NPS decisions and much agency behavior will reflect short-term political concerns. Unlike preservation-minded rangers and wardens, politicians and their appointees must necessarily be responsive to use demands. People who use the parks, as tourists, as recreationists, or as developers, are constituents. In short, U.S. national parks will be managed, often paved, for political purposes rather than preserved in a relatively natural condition.

Alternative Explanations

With my theoretical explanation for the recent divergence between the two park services, I do not suggest that other differences between the two agencies and the systems in which they operate are not important. However, I do argue that they alone are not enough to explain the recent behavior of the NPS and the CPS. Basing recommendations for reforms or future alterations on other explanations of agency behavior without considering the effects of goal consensus and political support would be ineffective.

GEOGRAPHIC AND DEMOGRAPHIC CONTEXT.   Could population differences between the two countries explain differences in behavior between the two park services? Canada and the United States are among the world's largest countries, second and fourth respectively. Canada comprises 3.85 million square miles, and the United States roughly 3.62 million. The United States protects as parks or wilderness areas about 10 percent of its land area, with roughly one-third of that total managed by the NPS. By the year 2000, Canada plans to protect 12 percent of its land area, with one-fourth of that managed by the CPS.[45] This will involve a huge expansion, since only about 5 percent is currently protected.

Demographically the countries are quite different. The U.S. population was 249 million in 1990, while the Canadian population totals 26.5 million. The U.S. population is also much more dispersed than that of Canada, with roughly 90 percent of Canadians living near the United States-Canada border. These differences are potentially important for parks. Canadian parks could be protected if they were in low-density areas far from the border, where they would presumably experience reduced visitation and demands.

In fact, many Canadian park sites are quite accessible to large population centers, while many U.S. parks are located in sparsely populated western areas. Figure 1-1 shows that most Canadian park sites are located near the border. The blackened circles represent areas managed by the CPS in the early 1980s. Many of the parks were established along railroad lines and highways in order to be accessible to the population. Figure 1-2 shows the distribution of sites in the U.S. system. The dots show how dispersed these units are and, indeed, how many are far from population centers. Demographic differences are important for the park systems, but, as these figures suggest, attributing different behavior by the agencies to these differences would be an oversimplification.

INCREASING CHALLENGES.   Disparities in population between the two countries lead to consideration of possible differences in visitation and other uses of the parks. One might hypothesize that NPS efforts at preservation have been altered more than those of the CPS because NPS units are subject to more challenging demands.

In many ways, the two agencies face similar demands and threats. The most obvious threat is overuse. The U.S. parks do host more visitors than do their Canadian counterparts. Currently, the NPS services roughly 58 million people at the fifty national parks alone in addition to other visits at monuments, battlefields, and other sites. At the same time, the thirty-four Canadian national parks entertain about 36 million visitors a year. Demand is indeed higher in absolute terms for the U.S. parks. In relative terms, however, the average number of visitors per park is quite comparable (1.16 million for U.S. and 1.06 million for Canadian). In addition, certain parks in each system share the brunt of visitor demand. Finally, both park systems have experienced tremendous growth in recent years. The number of visitors to U.S. parks has jumped by 70 percent in the past two decades but by more than 100 percent for Canadian parks during the same period.

In addition to visitation, parks in each system are subject to demands for a variety of uses. Both agencies have taken on increasing responsibilities

FIGURE 1-1. *Canadian National Park System*

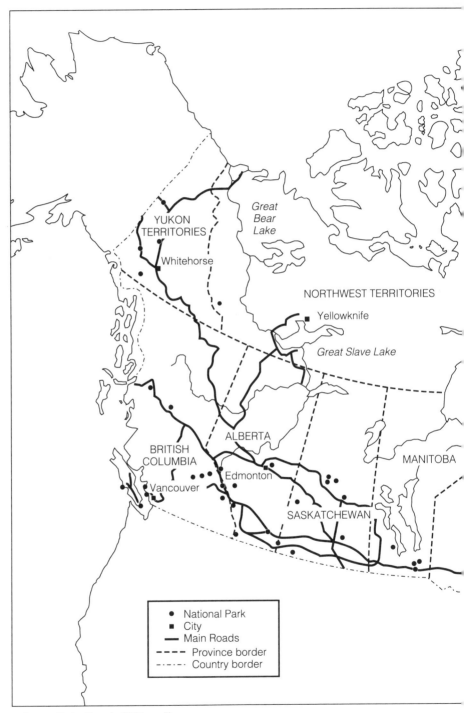

Source: Parks Canada (1985, p. 5).

FIGURE 1-2. *U.S. National Park System*

Source: U.S. National Park Service (1988c, p. 1.1).

with historic sites, battlefields, waterways, and other protected areas. Visitors to these sites are seeking increasingly diverse forms of recreation. For every hang glider and windsurfer in the United States, one can find a snowmobiler or cross-country skier in Canada. Parks throughout North America, indeed the world, are increasingly affected by external threats such as acid rain and global warming. Parks in both countries are subject to severe economic and commercial demands as well. Finally, both agencies have been forced to confront these numerous demands without increases in budgetary resources. The two park systems may experience different levels or intensity of demands, but similarities between the two countries in threats, uses, and challenges abound.

POTENTIAL PUBLIC COMMITMENT.   Finally, we might expect that commitment to the ideal of preserving natural areas depends on public support for environmental causes in general. Conceivably differences between the two countries on this issue could lead to differences in agency behavior. In fact, the two park service agencies enjoy similar levels of potential public support.

Public support for environmental issues exploded in the United States in the late 1960s and has since maintained such high levels that now "the public's commitment to a cleaner environment is stronger than ever."[46] The fact that the United States has yet to solve many of its environmental problems has more to do with inadequate government responses than with consistent public support.

Canadian sensitivity to these same concerns has usually trailed public attitudes in the United States. Possible reasons for the historical lag range from a Canadian sense of unlimited wilderness to the emphasis on economic growth to develop Canadian self-sufficiency. Recently, however, Canadian environmentalism has replicated that of the United States in events such as Earth Day and in developments such as the growth of environmental lobbies.

The current comparability of Canadian and U.S. public commitment to the environment is evident in a 1992 Gallup poll. The percent of respondents expressing a great deal or a fair amount of concern for environmental problems was 90 in Canada and 85 in the United States.[47]

Organization

The following chapters examine this proposed argument for the diverging behaviors of the two national park service agencies. Chapter 2 examines the argument in the context of the U.S. NPS. What caused the agency to change over the past fifteen years, and how did those changes affect behavior? Chapter 3 provides a comparable analysis of the CPS. Chapter 4 describes the effects of recent changes at the park level for both agencies. How do the discussed changes affect specific parks and those who spend time in them? Chapter 5 summarizes the findings, considers the broader implications of the study, and offers some recommendations for the two park services.

Many of the problems plaguing our national parks persist in spite of the efforts of trail-level bureaucrats in the NPS and the CPS. Perhaps more than others, they realize that our capacity for wonder is in jeopardy.

Chapter 2

Walking behind the Bulldozer

"Used to be that the super would walk along in front of the bulldozer, point to different spots, and say, 'Over there will be a campground, over there will be a picnic area.'"
—*Superintendent Harvey Wickware, Canyonlands National Park, 1991*

The National Park Service has changed over the past fifteen years. Now the superintendent often walks behind the bulldozer, the decisions having been made by someone else. To continue the analogy, the symbolic bulldozer rolls on, actions and decisions increasingly reflecting an emphasis on development and use of the parks for political purposes rather than for preservation of natural resources. These changes, which continue today, were neither welcomed nor desired by most National Park Service (NPS) employees. Why did they occur, and how have they affected agency behavior?

Until the late 1970s, the NPS enjoyed strong public support, maintained a decentralized structure, and developed a reputation for generally, with some noted exceptions, attempting to honor its mandate to allow access while preserving natural areas. Since then, conflicts over agency goals have become extremely visible and openly contentious. Political support for the NPS diminished even while the agency's responsibilities grew. As a result, external intervention into agency decisions increased, interagency disputes became more frequent, and authority for decisionmaking shifted to higher levels of the organization. Changes in organizational structure and operating conditions led to an increased concern with political incentives and a decreased ability to pursue the goal of preservation.

Background of the Agency

The NPS has a long and colorful history. Since its creation in 1916, the NPS has pursued an inherently difficult mission in an ambiguous structural setting. Until the 1980s the agency managed to maintain a precarious balance between conflicting demands and contradictory directives.

Defining the Mandate

On the evening I hiked the rim of the Grand Canyon, the Chicago Bulls were playing the Los Angeles Lakers for the NBA championship. My brother and I had a campsite but obviously no television. No problem. After setting up the tent, we strolled over to the El Tovar Hotel and found a bar with wide-screen color TV. National parks in the United States have not just campgrounds and hiking trails but also restaurants and bars. The expectation that parks will provide for both use and preservation dates to the mission statement of 1916 calling for the NPS to facilitate the "enjoyment" of the parks while leaving them in "unimpaired" condition. Managing parks to allow access while still preserving their natural condition is, as Grand Canyon's chief ranger told me, "the paramount issue" facing the National Park Service.

The dual mandate has roots that precede the agency. The main pressure to create a national park system that could preserve natural areas came from naturalists and groups such as the Sierra Club. Two events of the early twentieth century particularly motivated the preservationist impulse. First, despite the objections of John Muir, Congress decided in 1913 to flood Yosemite's Hetch Hetchy Valley. Second, in 1915, President Woodrow Wilson cut the size of Mount Olympus National Monument roughly in half to allow for lumber operations. Preservationists saw the "absence of a separate government bureau committed solely to their welfare and management . . . as the major threat to the future of the national parks."[1]

Preservationists got help, sometimes from unexpected sources. The western railroads, interested in transporting park visitors, became powerful allies. They and other economic interests made strong arguments about the need for parks by claiming that they kept tourist dollars in the country and provided relaxation. By stating that the park service would make parks accessible even while preserving them, the mandate avoided severely antagonizing conservationists, who advocated controlled use as the prime purpose of public lands.

Establishing Jurisdiction

National parks and national monuments existed before the National Park Service was created in 1916. In fact, confusion about who had jurisdiction over these places fueled the movement for a parks bureau. When the Antiquities Act of 1906 formalized the ability of the president to establish national monuments, Congress kept each national monument under the jurisdiction of its original bureau. This left thirteen with the Forest Service in the Department of Agriculture and fifteen with the Department of the Interior. The Department of the Interior (DOI) also managed the existing national parks (twelve by 1916), but Yellowstone remained under the jurisdiction of the War Department. Within Yellowstone itself, the Corps of Engineers was responsible for roads.

Official recognition of the lack of organization prompted consideration of the 1916 Organic Act. President William Howard Taft argued in 1912 that a parks bureau was "essential to the proper management of those wonderful manifestations of nature."[2] Members of Congress recognized the potential loss of the pristine nature of the parks because adequate protection was independent of electoral motivation. Senator George Vest, Democrat of Missouri, warned: "There are no votes in the Yellowstone Park for the Republican or Democratic party."[3] Thus, as Vest and others argued, the parks warranted their own managing agency.

Once the idea was born, the effort to establish a parks bureau received its leadership from two capable individuals, Steve Mather and Horace Albright. The leadership of these two men is important structurally because it ensured placement of the agency in the DOI. Mather was a wealthy outdoorsman who spent time hiking and mountaineering in western parks. After a visit to Yosemite in 1914, he was upset about the conditions there—"little protection, poor trails, and inadequate facilities for visitors." Mather expressed his anger in a letter to an old acquaintance from school, Secretary of the Interior Franklin Lane. According to Albright, Lane wrote back: "Dear Steve, if you don't like the way the national parks are being run, come on down to Washington and run them yourself."[4] When Mather showed up at Interior, Lane assigned Albright, a lawyer in the department, to help steer him through the inevitable red tape. The two coordinated campaigns, used newspaper and magazine publicity, and nurtured congressional support, all from within Interior.

Opposition to the proposals for the parks bureau also shaped the ultimate structure of the agency. The Forest Service and its leader, Gifford Pinchot, strongly opposed a new bureau, particularly one that would take lands from its jurisdiction. Others opposed the idea of another bureaucracy. The legislative leader of this group, Representative William Stafford, Republican of Wisconsin, had made a name for himself by opposing the creation of new agencies.

After numerous debates, the National Park Service was created in 1916. Because Mather and Albright worked out of Interior and because the NPS took over jurisdiction of Interior's monuments, the agency was placed in the Department of the Interior. As Representative Edward Taylor,

Democrat of Colorado, argued in the House, "We want to have a systematic distribution of improvements and expenditure of money under the service. If the monuments are left with the forest reserve, they may not be improved as rapidly or systematically in the way of roads and otherwise as if they were put under this park service all of them provided for in an orderly way."[5] Nevertheless, Forest Service monuments were not transferred into the new agency until 1933.

The Department of the Interior claims the title of "the Nation's principal conservation agency," and it is indeed responsible for most of the public lands, which constitute one-third of the country's total acreage. Interior's responsibilities range from endangered species (Fish and Wildlife Service) to strip mining (Office of Surface Mining Reclamation and Enforcement), water resources (Bureau of Reclamation), and Native American issues (Bureau of Indian Affairs). Interior is thus home to agencies whose own missions will inevitably lead to interagency conflicts over lands and waters in and surrounding the parks. The secretary of the department and the directors of the agencies within it are all political appointees and representatives of the president.

Compromises shaped several other structural arrangements affecting the new agency. First, the title itself indicated that the agency would be a service rather than a full-fledged bureau. This not only defused criticisms of excessive power, but it avoided making the Forest Service appear inferior. Second, proponents had seriously considered establishing an independent commission to oversee the agency. The commission would "develop policy, . . . make systematic inspections of parks, examine the effects of administration of the parks, and report to the Secretary of the Agency." The idea was abandoned because, as Albright put it, "We wanted to make it [the bill] as short and uncluttered as possible."[6] Third, the debate over final passage illuminated controversy over the secretary of the interior's direct power over the agency. The first amendment in the conference report takes away the ability of the Congress to provide for, "from time to time," certain employees of the agency and gives that power instead to the secretary. Mather had encouraged flexibility in remuneration of personnel so that the service could select "from among the most competent men." The conference committee did not give him and the NPS this power but instead "recognized the advisability of giving the secretary of the interior a certain degree of latitude in selecting other necessary clerical and expert assistants."[7] These structural arrangements all led to controversy in later years.

A Decentralized Structure

When I was working in Yosemite in 1978 I met Superintendent Les Arnberger. One night when he and his son were out on a camping trip, he came to my tent and borrowed some tools. Later the crew foreman seemed stunned that such an authority would stoop to talk to a lowly maintenance worker like myself. "Don't you know who he is?" he asked, concerned over what I might have said. "He's the boss, the big boss." Though I did not know it at the time, that exchange spoke volumes about the decentralization of the NPS.

The internal organization of the NPS was historically decentralized. When the one feature that all park units have in common is that they are different, only such an arrangement makes sense. Each park unit has its own distinct features, setting, and physical characteristics. Correspondingly, each park unit must be managed on its own terms. As a result, each park unit has its own budget, its own manager, and its own management plan. Park units come into existence for separate and distinct reasons. Because of their uniqueness, the national parks could never be quite as systematized as the national forests or other public land units. The seemingly innumerable categories of lands protected by the National Park Service give obvious evidence of the distinctive nature of park units. The NPS involves more than just what most people think of when they hear the term national park. In fact, the NPS manages twenty different categories of park units. The most prominent are national monuments, national historic sites, and national parks. Differences exist both within and between these categories. For instance, one important difference between monuments and parks is that monuments can be created by the president, while parks require an act of Congress.

At the end of the 1970s the NPS was the epitome of a decentralized agency. The agency managed nearly 280 units with an operations budget of roughly $400 million. The formal structure of the organization in 1979 is depicted in figure 2-1. As the diagram shows, the agency was divided into three levels: headquarters in Washington, D.C., ten regional offices, and the park units. Each park had its own superintendent. The superintendents have traditionally been the key decisionmakers in the organization. As the director of planning for the NPS told me, the traditional analogy for a superintendent is a ship captain. Others may own the property and determine the cargo, but once away from the dock (or in the field), the captain (or superintendent) makes the decisions.

Both personal accounts and systematic studies identify the superintendent as the ultimate authority. Former ranger and environmental godfather Edward Abbey attributed successful NPS operations to the behavior of field managers who are allowed to perform without higher-level intervention. After fifteen years as a smoke jumper in Grand Canyon, Stephen Pyne described the NPS as "not a system so much as an aggregate of baronies; the political philosophy is feudal, with each park administered by a superintendent possessed of great latitude." In his historical account

FIGURE 2-1. *Formal Structure of the NPS, 1979*

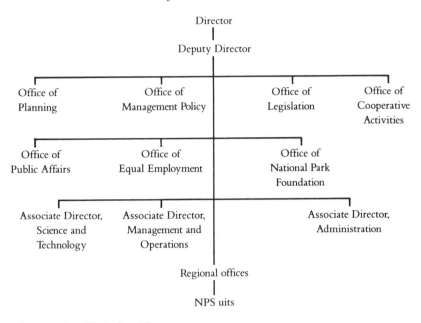

Source: Olsen (1985, chart 35).

of Yellowstone, outspoken critic Alston Chase used the term "Balkaniza-tion" to describe the agency with especially "the major parks . . . almost completely autonomous." Two noted political scientists, in their study of federal recreation budgets of the early 1970s, described the NPS as decen-tralized, with the budget estimates filtering up from the ground level.[8] Superintendents made nearly all of the day-to-day decisions affecting the parks.

To say that the NPS was decentralized in the 1970s should not be interpreted to mean that the agency was immune to politics. Indeed, po-litical intervention in park affairs has always been possible and, in some senses, enhanced by the decentralized system of unique units. As an agency within the executive branch, the NPS is obviously subject to influence from the president and his appointees. Further, the NPS is also subject to congressional control. The agency itself depends on Congress for its exis-tence. The most prominent units in the system are created by acts of Congress. Each national park is defined in a separate piece of legislation and receives its own line-item appropriation. This traditionally defined feature of distributive politics invites and tempts congressional involvement with specific units. Congress is also expected to oversee and to monitor the actions of public agencies. Members have traditionally had the potential to intervene in agency behavior.

Accommodating Competing Demands

The NPS has always been forced to deal with the potentially competing demands made explicit in the agency's mandate. An anecdote illustrates the point. While Bernard Shanks was serving as a ranger in Grand Teton, he learned that parklands were to be used as well as preserved. Theoretically the NPS does not allow cattle grazing on parkland. So when Shanks learned that 500 head of cattle belonging to Senator Cliff Hansen, Republican of Wyoming, a prominent member of the agency's oversight committee, were grazing in a Grand Teton pasture, he approached the chief ranger about removing them. The ranger made it very clear that Hansen's cows weren't going anywhere as long as Hansen was on the Senate Interior Committee. Demands such as the use of parkland for grazing and the ability of politicians like Hansen to make them have highlighted the shifting emphasis on either development or preservation.

Nevertheless, the National Park Service built a reputation that included at least some attention to preservation. The NPS was designated as separate from the Forest Service largely to ensure that preservation, rather than the multiple-use concept that governed the administration of the forests, was its main objective. Until the 1970s the NPS usually emphasized some preservation of natural conditions even while allowing access. A 1971 comparison of U.S. public land agencies noted: "Give several agencies a single problem to solve and each will respond with its own solution . . . the National Park Service is apt to prescribe preservation." During the 1970s the preservation emphasis was reinforced by the environmental awakening throughout the United States. Although it was occasionally criticized, the NPS was rarely described as being "captured" by development interests and instead more often accused of "locking up" the wilderness. By the end of the 1970s, the agency's historian pronounced, "The Park Service will continue to move in the direction of preservation, but only slowly." An outside expert concurred in 1978: "The Park Service . . . has gradually come to favor park preservation emphases as more important than facilitating recreational uses."[9]

At the end of the 1970s the NPS was perceived as an agency that managed to pursue a difficult mission under less than ideal conditions. Though it never received as much funding or support as some other public land agencies, the NPS was still depicted as able to "muddle through."[10] Furthermore, the employees of the NPS were generally upbeat; 84 percent in one survey expressed optimism about the agency's future.[11] The perceptions of observers and of agency employees were about to change.

Changes in the Political Environment

During the 1980s, two political factors affecting the NPS changed dramatically. First, inherent tensions over the potentially contradictory goals of the agency came into open conflict. Second, political support for the NPS, through external structure and budgetary resources, diminished. Thus when the NPS needed consistent leadership and support the most, it received just the opposite.

Absence of Consensus

During the 1980s the NPS was subjected to severe demands on both extremes of the perceived dichotomy between preservation and use. One example clearly illustrates the conflicting demands. In an interview two months after taking office, new Secretary of the Interior James Watt admitted, "If I err, I'm going to be erring on the people side." In a speech shortly after his appointment in 1985, NPS director William Penn Mott pronounced, "We must err on the side of preservation."[12] The tension between these points of view was exacerbated by the development-oriented preferences of the Reagan administration, the increasing efficacy of environmental interest groups, and the increased awareness of and publicity surrounding the challenges facing the agency.

ERRING ON THE SIDE OF USE. The views of the Reagan administration on national parks were stated early. In a 1981 speech to park concessionaires, Watt said: "It is time for a new beginning and the private enterprise system must be looked to for rejuvenation and enthusiasm as we try to make our parks more accessible and usable for the people."[13] As point man for the administration on public lands issues, Watt issued directives for all the agencies in Interior and set the tone for other departments. Mott's stated preference for preservation was heavily outweighed within the administration as other appointees mirrored Watt's intentions and often even his style.

Examples of the Reagan administration's priority for making the parks "usable" abound. Watt and his assistant secretary for fish, wildlife, and parks, G. Ray Arnett, tried to lease coal mining in Chaco Canyon, attempted to reverse the ban on strip mining near Bryce, opened up Lassen to snowmobile use, and ordered career personnel transferred out of Glacier Bay for attempting to protect the endangered humpback whale by imposing restrictions on tour boats. Despite strong public opposition, they tried to change commercial fishing regulations in the Everglades. Arnett, an avid hunter and founder of a group of trophy hunters, wanted to allow hunting of mountain lions near Carlsbad Caverns. Other Interior officials forbade the NPS to strongly oppose a proposed nuclear waste dump site at the edge of Canyonlands.

Even after Watt was driven from office following controversial actions and insensitive remarks, the preferences of the Reagan and subsequently the Bush administrations did not change. Watt's successors, William Clark and Donald Hodel, were less abrasive, but their policies consistently favored development. When Arnett left to become executive director of the National Rifle Association, his successor, William Horn, instigated perhaps as many disputes with the NPS as had Arnett. Horn battled the NPS over specific decisions, personnel evaluations, and even policy statements. For example, when the NPS director, amending the agency's management manual, tried to emphasize avoiding impairment of natural conditions, Horn objected, calling impairment a "highly subjective" term.[14]

The situation changed little after Reagan left office. The Bush administration, through the offices of Interior secretary Manuel Lujan, Jr., endorsed programs similar to those of the Reagan years. Lujan's attitude is typified by something that occurred in 1990. A religious group known as the Church Universal and Triumphant settled on the border of Yellowstone. They moved in with buildings, fences, and automatic weapons and drilled for geothermal wells. When Yellowstone officials described the development as one of the greatest threats to the park, Lujan asked to meet with the church's leader to work out a solution. Following the meeting, Lujan called the developers no threat to the park. Park officials were stunned.

The Reagan and Bush administrations were supported by outspoken groups that were particularly active in western states. Advocates for reduced federal restrictions on the public lands in the West were prominent as members of the sagebrush rebellion throughout the 1980s. The sagebrush rebellion provided support and personnel for the Reagan administration through both terms. More recently demands for economic use of the parks have been proposed and applauded by advocates of the Wise Use movement. This movement unites such groups as the National Rifle Association, the National Cattlemen's Association, Exxon USA, and others under the banner of property rights. They advocate expanding private use of national parks by allowing development, timber harvesting, grazing, and mining. The agenda of this group is directly counter to the goals of preservation. In fact, the Wise Use Mission 2010 statement calls for opposition to any actions "inimical to the mandate of Congress for 'public use and enjoyment' in the National Park Act of 1916."[15] This group had a receptive audience in the Bush administration for many of its positions.

TABLE 2-1. *Membership in Relevant Interest Groups, 1979–90*

Year	National Parks and Conservation Association	Sierra Club	Wilderness Society
1979	35,000	183,000	55,000
1980	30,000	199,000	50,000
1981	33,000	225,000	50,000
1982	45,000	310,000	60,000
1983	32,000	246,000	54,000
1984	45,000	350,000	110,000
1985	60,000	350,000	140,000
1986	60,000	370,000	150,000
1987	65,000	416,000	190,000
1988	100,000	416,000	225,000
1989	100,000	500,000	315,000
1990	200,000	565,000	390,000

Sources: *Encyclopedia of Associations*. The year depicted is two years before the date of the volume to allow for compilation and publication time. The figures for 1983 and 1986 are estimated from Ingram and Mann (1989, p. 140).

ERRING ON THE SIDE OF PRESERVATION.   While the executive branch pushed the National Park Service in one direction, environmental interest groups, often acting through members of Congress, pushed the other. Wilderness areas, and parks in particular, have inspired organized support since the late nineteenth century. The Sierra Club, founded in 1892, grew up around John Muir's interest in Yosemite and other areas of the Sierra Nevadas. The Sierra Club and other organized interest groups, particularly the National Parks and Conservation Association (NPCA) and the Wilderness Society, have become increasingly involved in parks and park policies. These groups grew in strength during the environmental awakening of the late 1960s and early 1970s. But the most dramatic increases in their growth have occurred in the past fifteen years. Table 2-1 portrays the phenomenal growth of membership in these groups in the 1980s.

The relationship between growth of environmental interest groups and demands on agency behavior is recursive. In other words, the two phenomena reinforce each other rather than being simply causal in one direction. Understanding increasing membership in response to agency visibility is relatively straightforward. By many accounts, the membership rolls of environmental interest groups swelled in response to the highly visible behavior of Reagan and Watt. The other causal direction is a bit more complex. Interest groups must be concerned with attracting and maintaining members as well as pursuing policy preferences. After all, the pursuit of goals is not possible without membership. Interest groups thus publicize not only their activities but also the need for their activities. Greenpeace not only takes photographers along on anti-whaling missions, but the group publishes articles on the depletion of the whale population. Groups interested in the parks have thus become quite vocal in their demands.

Most of these environmental groups, particularly the NPCA, have strong preferences for the preservation aspect of the NPS mission. The NPCA was established just three years after the National Park Service at the suggestion of the NPS director, Steve Mather. NPCA describes itself as "America's only private nonprofit citizen organization dedicated to the protection and preservation of our national parks." The actions and statements of the NPCA have become much more critical of the NPS in recent years. In an interview with me, President Paul Pritchard expressed his criticism of the agency: "We have institutionalized the things that don't work rather than the things that do work." The NPCA's membership solicitations have an increasingly urgent tone: "You may never have another chance like this one. If you want to see America's National Park System survive. . . ."[16] Editorials in the NPCA magazine match this dire tone. Pritchard is the author of numerous articles in the NPCA magazine and in other magazines such as *National Geographic.*

These environmental groups attempt to influence the NPS in general and in individual cases. Increasing numbers of policy proposals and reviews of agency behavior are available for general consumption. The NPCA, for example, is now publishing what seems to be an annual series on threats to the parks. The 1991 report *Parks in Peril* was followed in 1992 by *Parks in Peril: The Race against Time Continues.* Not surprisingly, these documents are released to and touted by the press before becoming available to the public. The NPCA and other groups have also published numerous compelling books on the parks in the 1980s. Many groups have lobbied and testified on issues such as fee increases and concessions policies that concern the NPS. The tone of these groups has become decidedly more aggressive, political, and audible.

Finally, these groups are increasingly involved in specific agency policy decisions. In many cases environmental interest groups have worked with NPS employees to fight attempts to develop and commercialize the parks. Still, those relations have occasionally been strained. One superintendent, who asked that even the situation not be identified, described negotiations with outside developers that were aborted because of publicity generated by environmental groups. An example reported in the press involved confusion among Russian negotiators over the proposed Beringian Heritage International Park when both the NPS and environmental groups released bilingual proposals.[17]

Challenges "Greater than at any Time"

On the border of Zion National Park in southern Utah is the little town of Springdale. The northern edge of the town is just outside the main entrance to the park. Not much is there now except a general store and a few houses. The store is visible across the Virgin River from the major campground (Watchman) in the park. Within a few years, however, the view will be quite different. California-based developers are planning to build a four-story, big-screen movie theater complex complete with parking lot and a lodge with 120 rooms. Though park officials have fought the plans, town leaders have approved them. The theater will show a film of park attractions for those who do not want to enter park grounds; this has gained for Zion the nickname Couch Potato Park. The issue is no laughing matter to those who love Zion. Park superintendent Larry Wiese pointed out the site to me, shook his head, and described the situation as "a tragedy."

In recent years the threats facing national parklands have become more numerous and more apparent. Some of these threats have existed for decades but have only recently received serious attention. As the agency's own historian says, the NPS "faces challenges greater than at any time in its history."[18] A comprehensive document entitled *State of the Parks*, published by the NPS in 1980, identified and classified numerous threats, both internal and external, to the parks. In a survey several years later, the General Accounting Office (GAO) concluded that 80 percent of the threats were unresolved and 43 percent undocumented.[19] Public attention to these threats has increased, thereby intensifying debates over NPS action (or inaction).

INTERNAL THREATS.   The parks have always faced certain threats endogenous to the nature of parks themselves, the most obvious being potential overuse by visitors. As early as 1926 the agency's official report cited "indisputable evidence of the constantly increasing popularity of the national parks."[20] In the year following its establishment, the NPS serviced 488,268 visitors. By 1960 total visits had grown to 135 million, and ten years later the figure stood at 200 million.[21] The growth in the 1980s, up to more than 300 million, is shown in table 2-2. Similarly, recreational visits rose steadily, except for a brief lag in 1985, between 1980 and 1992.

As dramatic as these figures are, the problems associated with increased visitation are intensified by several other factors. First, many of the visitors are confined to certain parks. In 1989, 55 percent of the visits to national parks were confined to 15 percent of the parks. This concentration has necessitated the installation of a campground reservation system at the fourteen busiest parks in the system. Second, even within the

TABLE 2-2. *Total Visits to NPS Sites, 1980–88*

Year	Total
1980	294,582,280
1981	327,348,000
1982	331,455,176
1983	337,947,177
1984	328,392,400
1985	347,221,086
1986	352,155,500
1987	370,982,172
1988	371,489,100

Source: U.S. Department of the Interior (1990a, p. NPS-12).

individual parks, visitors are often clustered in specific locations, such as Yosemite Valley, thus accelerating the deterioration of certain areas. Third, most of the visits come in "surges," as field managers term them, when the park is inundated in certain months of the year. These factors have contributed to such a serious deterioration of the infrastructure in the parks that NPS director James Ridenour called it his "biggest shock" upon entering office in 1989. For example, one-third of the 7,900 miles of roads in the parks need repairs, at a total estimated cost of $1.5 billion.[22]

Together these issues make visitor impact an increasing threat. This threat has been apparent to park observers for decades. Since its inception the National Park Service has precariously steered the parks between enough visitation to promote the concept of the parks and more visitation than the parks can handle. In stages such as the Mission 66 period, a ten-year program initiated in 1956 to develop the parks, the agency was extremely aggressive in the pursuit of visitors. As early as 1953, Bernard Devoto described how deterioration of parks "has been accelerated by the enormous increase in visitors," as he demanded, "Let's Close the National Parks."[23] At that time Devoto's was a lonely voice in the wilderness. Today the chorus is resonant. The effects of visitation have been increasingly publicized in the 1980s. From a *National Geographic* article, "Are We Loving it to Death?" in 1978 and *Newsweek*'s "The Grand Illusion" in 1986 to a *Life* cover story on "The Fragile Balance" in 1991, the popular press has been extremely vigilant in publicizing the effects of visitation. A comment from a 1991 series in the *Chicago Tribune* is typical: "An unprecedented number of tourists from the U.S. and the world are flocking into the national parks, forests, and public lands, their love of the outdoors becoming a threat to the treasures that attract them."[24]

A threat that is related to visitor impact is identified in *State of the Parks* as aesthetic degradation. Not only are the parks handling greater numbers of visitors, but the activities of visitors have changed. As the American population ages, more visitors have become "windshield tourists," preferring to see the sights from behind the wheel of motor vehicles, often in the form of huge trailers or recreational vehicles. For example, 70 percent of the visitors in 1960 spent two to five days at one spot, but 55 percent of the visitors in 1990 spent six hours or less in one spot.[25] This translates to more roads and fewer hiking trails. More and more visitors want not only amenities like showers but also recreational activities such as skiing, golfing, and motorized boating that affect the land more than traditional activities such as hiking and camping. Individual situations, such as increased lodging at Yosemite and the use of snowmobiles at Yellowstone, have been highly publicized in books and articles.

An increase in visitors inevitably translates to an increase in the crime rate. These days the jail at Yosemite is, as one ranger told me, always busy. But crime is hardly limited to this paradise. The General Accounting Office estimates that the annual average number of criminal offenses in the parks is more than 40,000, including about 6,500 major crimes a year. The NPS reports that vandalism costs several hundred million dollars a year. Crime and subsequent NPS responses, such as rangers carrying .357 Magnums, are not what most visitors hope to see when they come to the parks.

A third internal threat identified by the NPS and increasingly publicized in the past decade is the removal of resources from the parks. Specific problems range from specimen collecting and mineral extracting to poaching. Extracting resources from areas in and near national parks has always been controversial. Mining efforts, particularly with oil and coal, have intensified lately as other sources of fossil fuels dwindle and as potential supplies are found on parklands. For example, the U.S. Geological Survey has identified twenty-five NPS units with underlying coal fields. Another natural resource that tempts extraction and commands considerable media attention is wildlife. Poaching and the destruction of big game such as bears, cougar, and elk have been the subject of stories in the press in recent years.

A fourth internal threat admitted by the NPS and frequently cited by critics of the agency is in the category of park operations. The NPS itself has identified examples such as misplaced roads, misuse of biocides, and suppression of natural fires. The last issue especially generated controversy after the Yellowstone fires of 1988. Other NPS operations that have received critical coverage include wildlife control, monitoring of waste dumping, and ecosystem management.

The category of park operations has received particularly acute public attention since Alston Chase published his controversial *Playing God in Yellowstone* in 1986. In this highly critical account, Chase chastised the

NPS for its management of the world's oldest national park. Chase asserted that "over the last seventy years nearly every conceivable mistake that could be made in wildlife management has been made by the Park Service in Yellowstone."[26] Not surprisingly, NPS officials are extremely sensitive to criticism of this sort. When I asked one Yellowstone manager about Chase, he responded brusquely, "Everybody's entitled to his opinion." These days, it seems, everyone has an opinion about park operations, and many of them make it into print.

EXTERNAL THREATS.   Concern over threats to national parks has dramatically increased in the past decade as a result of greater awareness of previously unrecognized external issues. State of the Parks quantified the external danger: "More than 50 percent of the reported threats [to parks] were attributed to sources and activities located *external* to the parks."[27] Since then two things have happened. First, the number of threats has increased. As one superintendent told me, if a manager still concentrates exclusively on internal balancing of use versus preservation, the park will suffer severely from unaddressed external issues. Second, awareness of these threats has increased. In the past decade, numerous reports have publicized external threats in the following categories identified by the agency.

Fully 16 percent of the reported threats in the *State of the Parks* survey concerned air pollution. Problems concerning contaminants, visibility, and acid rain have all increased since that survey. Air pollution has been documented at parks from the Colorado Plateau to the Blue Ridge Mountains. The most affected park in the system is Shenandoah. As air pollution has increased recently, particularly over the past ten years, average visibility in the park has declined from 150 to 30 kilometers. In addition, pollution kills plant life, and acid rain destroys aquatic resources. Much of the pollution, according to sophisticated monitoring set up by the park, is generated by nearby coal-fired power plants. Many of these plants have only recently received operating permits, fifteen of them since 1987.

Water pollution derives from toxic chemicals, unnatural flooding, oil spills, and other external sources. The park most affected by water problems is currently the Everglades, where fresh water flowing into the park has been diverted and replaced by agriculturally contaminated water sources.

One of the more publicized sources of water contamination is hydroelectric power generation on a river outside a park's boundaries. Dams and hydroelectric power have affected parks and water within parks for decades. For example, the impact of the Glen Canyon Dam on the Grand Canyon has been well documented. Proposed dams such as one on Copper Creek near North Cascades and two on the Virgin River outside of Zion are likely to affect those parks in coming years.

Wild and scenic rivers under the jurisdiction of the NPS are often affected directly by external sources of pollution. The Chattahoochee River, for example, is "seriously impacted by raw sewage and chemical runoff from adjacent lands."[28] This category also includes less known problems such as the underwater devastation of the coral reefs in Virgin Islands National Park caused by the anchors of cruise ships. Geothermal drilling just outside of park boundaries affects water quality in Lassen, Grand Teton, and Yellowstone.

A third category of threat, encroachment, includes nonindigenous plants and animals, unwanted noise, and fires caused by humans. Some of these issues are more obvious and more publicized than others. Countless reports of aircraft buzzing whitewater rafters in the Grand Canyon have led to contentious battles over flight restrictions. Other issues attracting media attention involve exotic species as well as the reintroduction of formerly indigenous species such as wolves that have been removed from parks. A classic example of the impact of nonindigenous species involves the controversial Fishing Bridge area of Yellowstone. Cutthroat trout were nearly crowded out by exotic species, longnose sucker and redside shiner, that were brought in for bait. The NPS has since banned fishing from the bridge, and the trout population is staging a comeback, but the bears that used to feed on these fish have not returned. The fight over this area has been contentious and visible.

Finally, aesthetic degradation includes even more external threats than internal. These include a wide range of issues that threaten enjoyment of a park, issues such as land development outside the park and urban encroachment. Land development is going on near many parks and historic sites. Any visitor driving into Olympic, for example, will undoubtedly notice the clear-cut timberlands surrounding the park. Many historic sites, particularly Civil War battlefields, are threatened by the development of residential or commercial complexes just outside park borders. Situations involving Antietam and Manassas have been heavily publicized. The Springdale theater plans near Zion that were mentioned earlier have stirred opposition from well-established environmental groups and from celebrities such as Robert Redford.

In general, the threats facing the national parks have become more numerous, more complex, more apparent, and more publicized in the past fifteen years. These developments have intensified inherent tensions among NPS goals. The intense debate over agency actions and the dramatic challenges facing the NPS have contributed to a lack of consensus over the mandated goal of preserving natural conditions. Official policies and associated documents facilitated decreasing consensus with, as legal scholars said, "a remarkable lack of elaboration and clarification of legislative intent."[29] Observers and analysts contributed to the perception of virtually unresolvable conflicts by describing the NPS mandate as a virtual "Mission Impossible." By the end of the 1980s many within the agency were begging for "leadership that is capable of enunciating and implementing clear and compelling goals for parks policy and Park Service management."[30]

Diminution of Political Support

In the agency's own vision for its role in the next century, the NPS reviewed the past two decades by admitting that it "has lost the credibility and capability it must possess in order to play a more proactive role in charting its own course, in defining and defending its core mission."[31] Losses of credibility and capability are manifestations of the reduction in political support that the NPS has experienced since 1980. The agency no longer enjoys the respect of other people in government (credibility) or the resources (capability) it needs to implement its programs.

AGENCY PLACEMENT.   Since its establishment, the NPS has been housed in the Department of Interior. The DOI is supposed to manage public territory according to "the wisest use of our land."[32] Because the DOI comprises a wide mix of agencies, determining exactly what "wisest use" means is difficult. Many public lands are managed by agencies outside of the DOI. National forests, for instance, are administered by the Forest Service, which is under the jurisdiction of the Department of Agriculture. The public lands have been set aside for divergent purposes and widely differing constituencies. Different missions lead inevitably to disputes for agencies managing adjacent, sometimes even overlapping lands. Those disputes increased in number and in intensity during the 1980s.

To some extent, disputes are inevitable. The federal government owns more than 700 million acres of U.S. land. The DOI controls nearly 500 million acres, with the Bureau of Land Management (BLM) overseeing more than 300 million of those. Of the remainder, the U.S. Forest Service manages roughly 190 million acres. Many of the BLM and Forest Service lands border national parks. These lands are managed for purposes other than providing buffer zones for the parks, since the mandates of these two agencies are quite different from that of the NPS. National forests are to be managed for multiple uses including recreation and logging. BLM lands have served economic interests for so long that the agency is commonly referred to as the Bureau of Livestock and Mining.[33] Operations of both agencies have resulted in differences of opinion with the NPS for decades.

The number and intensity of intergovernmental disputes involving the NPS greatly increased in the 1980s. First, while the Department of the

Interior had never been completely consistent in its directives for public land management, past secretaries had usually expressed at least rhetorical support for goals consistent with those of the NPS.[34] Reagan's public lands program emphasized two goals that were heartily endorsed by DOI leadership: privatize as much land as possible and lease the resources on the rest. Preservation was a lost cause for the Reagan administration. NPS careerists espousing such a goal were shown little respect or deference by DOI officials and other agency heads who adopted the administration's line.

Second, other public land agencies, headed by appointees with views similar to those of Secretary Watt, aggressively applied the Reagan program to lands adjacent to or near parklands. As one observer noted, the DOI often "weaken[ed] park protection by permitting sister agencies to disregard or minimize any legal priority for park protection."[35] Because external factors can affect the parks, even if these agencies respected the NPS, their own actions could lead to intensified disputes.

Third, cuts in operating and land acquisition funds significantly impaired the ability of the NPS to redress the consequences of developments allowed by other agencies on adjacent lands. NPS officials could protect the land only by objecting to the actions of other agencies rather than by dealing solely with the effects on parklands.

Consider one single river as an illustration of how easily interagency disputes can be generated. The Elwha River in Olympic National Park is managed by the NPS. Because the Elwha is surrounded by forests, it is subject to siltation and debris from timber harvesting on lands managed by the Forest Service. If mining takes place on nearby lands managed by the BLM, the river could be affected by pollution and tailings. Agricultural use of neighboring lands, regulated by the Department of Agriculture, can release pesticides and herbicides into the waterway. The NPS is currently fighting the Federal Energy Regulatory Commission over the Glines Canyon Dam on the Elwha. As all of these agencies become less concerned about the NPS mandate, interagency interactions and disputes will inevitably increase.

Interagency disputes within and outside of the DOI during the 1980s were abundant. The BLM, under the leadership of Robert Burford, the former leader of the sagebrush rebellion, clashed with the NPS over geothermal development near Yellowstone and oil development adjacent to Arches. The NPS received little political support from the DOI for dealing with its sister agency. Referring to external threats facing the parks, Watt told the NPS director, "You're on your own."[36] Outside of the DOI, the Forest Service, under ex-lumberman John Crowell, emphasized the logging aspect of the Forest Service's mandate to the point of inspiring serious disagreements with the NPS over lands near Redwood, Yellowstone, and

Glacier, among others. Further, since the confrontational styles of appointees Watt and Environmental Protection Agency administrator Anne Gorsuch attracted so much attention to public lands, many of these disputes were magnified and intensified.

Intergovernmental disputes did not disappear with the end of the Reagan administration. The Bush administration openly maintained the user-friendly attitude toward public lands. Furthermore, though organizations like the Forest Service claim they are becoming more conscious of the preservation mandate, many observers remain skeptical. The agency's sacking of internal reformers like John Mumma in 1991 for criticizing policies favoring timber companies reveals what many see as the true policy preferences of top administrators.[37] Those preferences will inevitably lead to disagreements over lands neighboring or affecting the national parks. For example, recent NPS policies concerning problems at Grand Canyon necessitated coordination with the Environmental Protection Agency for air pollution and with the Bureau of Reclamation for water erosion. As a result of the inevitability of continuing intergovernmental disputes, personnel within the agency made the need to "strengthen interagency coordination" official policy in the agency's *Twelve-Point Plan* for long-range strategy.[38]

DIMINISHED FINANCIAL CAPABILITY.   The diminution of political support for the NPS in the 1980s included reduced financial assistance. Factors that contributed to diminished financing included Reagan–era budget policies, congressional arrangements for revenues from increased entrance fees, and park concession contracts.

The Reagan administration came into office promising severe cuts in government spending. While somewhat inconsistent in terms of success, Budget director David Stockman and his team did have an effect on the parks. Table 2-3 shows total appropriations to the NPS throughout the 1980s. The congressional budget process works several years in advance. Spending in fiscal years 1982 and 1983, for instance, reflects the influence of the Carter administration. After that, the decline in appropriations to the agency until the end of the decade is obvious. These amounts are not adjusted for inflation, so the decline is even more severe than the numbers indicate. Further, the appropriations included substantial increases made by Congress over what the administration had requested for the agency. In real terms, the NPS operating budget has remained flat since 1983.

These budget figures had significant effects on the NPS for several reasons. First, while NPS responsibilities increased dramatically with new units, more visitors, and greater threats, agency funds were spread even more thinly than they had been before. Second, the money that was made available was concentrated on construction and maintenance. Funds for

TABLE 2-3. *Total Appropriations to the NPS, 1982–90*
Thousands of dollars

Fiscal year	Budget authority	Appropriations
1982	802,355	802,355
1983	1,106,846	1,106,846
1984	948,390	962,390
1985	975,327	1,003,327
1986	876,541	886,398
1987	962,717	975,217
1988	943,349	974,349
1989	1,075,088	1,122,088
1990	1,131,793	1,143,793

Source: Unpublished U.S. NPS document on appropriations, August 2, 1991.

park protection and historic preservation were cut to allow for capital developments. Third, Secretary Watt refused to spend money already available in the Land and Water Conservation Fund on new parks and parkland acquisitions, including 423,000 private acres the NPS had identified within park borders. As a result the price of the land skyrocketed. Watt maintained his stance even in the face of congressional appropriations for purchases.

A second financial issue is less well understood and thus requires more explanation. In 1986 Congress raised entrance fees, often from $1 to $5, throughout the park system. The increases climaxed years of effort by the Reagan administration to stop "subsidizing campers."[39] The crucial result for the NPS was that it did not benefit financially from the increases.

Several arguments were offered to support fee increases. First, entrance fees at the sixty-three units charging admission in 1982 had not changed in ten years. During that time the use of the parks and the expenses required to operate them had changed. For example, a 1982 GAO document reported that operation and maintenance costs rose 296 percent between 1971 and 1981. Increased fees would, the Reagan people testified, bring the user share of NPS operating costs back up to the 11 percent it had been when fees were frozen.[40]

Second, increases in entrance fees would generate greater revenues from the parks. This argument was based on two assumptions. Demand for park use is generally thought to be inelastic. Proponents argued that higher fees would not discourage visitation, and they used GAO studies as well as examples of increases at state parks to buttress their argument. The second assumption for increased revenue involved the motivation for park personnel to collect fees. The proposals for increased fees promised higher appropriations for the parks that collected more revenue. Presumably, this "capitalism at work in the park service," as Senator Dale Bumpers, Democrat of Arkansas, described it, would encourage effective collection efforts.[41]

Third, higher revenue would translate to more appropriations for the parks in general. Administration representatives testified repeatedly that the increased revenue would be used to supplement, rather than to supplant, normal NPS appropriations. Without unanimously endorsing any single plan, proponents of the fee increases based their argument on assertions that the revenue would bypass the normal appropriations procedure on the way to the NPS. Though some witnesses expressed skepticism about this promise, the need for greater funds available to the NPS was recognized broadly by environmental groups, the recreation industry, the recreational vehicle lobby, and the concessions industry.[42] The NPS, in its annual reports of 1986 and 1987, promised that the money collected would be used within the parks.

The contention that fee revenues would supplement normal appropriations was the argument that received the most consideration during the legislative process. The proposition that Congress would pass along increased revenues while resisting the temptation to use them for programs other than those in the parks aroused some wariness and suspicion. Even the supportive Senator Malcolm Wallop, Republican of Wyoming, admitted at one point, "I don't trust us." One witness testified that the promised bypass of the appropriations process was illegal and had been ruled so on the day before the House acted on the bill. This action, as well as general disbelief that increased revenues would be spent in the parks, inspired the president of the Sierra Club to declare his organization's opposition to fee increases.[43]

A second category of arguments against fee increases received only passing mention in the debates. These arguments hinted at changes in congressional control of the NPS and were based on the fear of excessive emphasis on visitor totals. The Sierra Club, Senator Bill Bradley, Democrat of New Jersey, and other witnesses expressed concern that concentration on the total number of visitors would lead to a slippery slope differentiating popular and unpopular parks. Administration representatives did not deny the importance of visitation figures, having already argued that such emphasis was "not inappropriate." Perhaps surprisingly, only a few questioned the legitimacy of raising the price to see America's public lands. When the agency had been created in 1916, Steve Mather had said: "I think the principle of having the parks free is the proper one, and we hope the time will come when we can make them free to motorists as well as to others."[44] One former NPS official, Howard Chapman, echoed Mather by saying it was "not appropriate to charge at a National Shrine."[45] He received little support among members of Congress except when they referred to parklands within their specific purview. Exemptions were requested for Alaska parks by Frank Murkowski (Republican of Alaska), for Hot Springs by Dale Bumpers (Democrat of Arkansas), for Acadia by William Cohen (Republican of Maine) and George Mitchell (Democrat of Maine), for Mt. Rushmore by Larry Pressler (Republican of South Dakota), for Cuyahoga by Howard Metzenbaum (Democrat of Ohio), and for the Statue of Liberty by Bill Bradley (Democrat of New Jersey) and Daniel Patrick Moynihan (Democrat of New York).[46] They argued that their accessibility to urban areas made fee collection more difficult, that the park's authorizing statutes prohibited entrance fees, and occasionally that it was undesirable to put a price on a national treasure.

Formal authorization of entrance fee increases took years. By 1986 and 1987, the huge federal deficits dominated other domestic concerns to such an extent that the suggestion of greater revenue from higher fees was overwhelming. The increasing popularity of the parks made the prospect of more appropriations available for operating expenses even more tempting than it had been before. Finally, proponents of increased fees played hardball with the budget. The administration's proposed 1986 budget promised an increase in the NPS operating budget of $32 million only if Congress enacted higher fees. Otherwise the NPS operating budget would actually be cut by $28 million.

Congress responded with increased entrance fees beginning in 1987. Automobile entrance fees were raised to $5 at many parks and to $10 at Yellowstone, Grand Teton, and Grand Canyon. Fees charged per person were raised at even more parks. Fee revenues were to be made subject to the appropriations process, but, "To the extent feasible, such funds should be used for purposes which are directly related to the activities which generated the funds."[47] Revenues passed back to the NPS were to be allocated according to the following formula: 10 percent by need, 40 percent by operating expenses, and 50 percent by proportion of fees collected. Numerous park areas such as Independence Hall and the Statue of Liberty were exempted from the fee policy. Urban parks with public access at multiple locations were also excused from charging admission fees.

Higher entrance fees did lead to increased NPS revenue, but not to corresponding increases in agency appropriations. Increases in agency revenue were greater than could be explained simply by the increases in visitor totals.[48] The debate over whether or not increased revenues would accrue directly to the agency was settled by the Conference Committee on Appropriations in the fall of 1986. In their report, the conferees noted: "To simplify the collecting, accounting, and utilization of fee receipts, the language [of the appropriations bill] provides that the fee receipts shall be deposited directly into the General Fund."[49] In the General Fund those revenues became available for everything from deficit reduction to park services. Indeed, the prohibition on the "bypass" asserted by opponents of fee increases in the debates was formalized by the statutory language of the legislation.

As a result, the promised increase in funds for the agency that was to accrue from higher fees never did materialize. Levels of appropriations for the NPS are affected by many variables, so I offer two perspectives to show what has happened to NPS appropriations. The first is historical. Figure 2-2 shows increased appropriation levels for the Forest Service and Grazing Service after well-publicized fee increases on their lands. Unlike the NPS, both agencies enjoyed periods of budget growth after fee increases. The NPS curve is noticeably flat.

Second, comparison with other public land agencies during the same time period is useful. Figure 2-3 shows the money available to the Park Service in comparison with other public land agencies that did not increase entrance fees at the same time as the NPS. If NPS outlays had increased as a result of the policy change, their outlay levels would show increases where those of the other agencies did not. In fact, no relationship between increased fees and increased appropriations for the NPS is apparent. While one could argue that much of the budget determination preceded the years in which entrance fees were established, proponents of the policy change had promised immediate results. The absence of any correlation between fees and appropriations is certainly not what was promised in the debates.

The third financial issue concerns concessions. The national parks have had concessionaires since the beginning. Indeed, five hotels and two lunch stations existed in Yellowstone even before the NPS was created. The Park Service has encouraged and relied on concessionaires to attract and to entertain visitors. The concessionaires operated with considerable autonomy until 1965.

The Concessions Policy Act of 1965 established the following relationship.[50] A concessionaire for an individual park obtains a long-term (up to thirty-year) monopoly contract, which protects almost unconditionally against loss, underwrites park improvements, and grants preferential rights for renewal when the contract expires. In addition, the concessionaire gains a "possessory interest" (all but legal title) in improvements, which makes replacing the concession holder difficult even when contracts expire. Public input is rarely sought or heard when contracts expire, a point on which concessionaires historically insisted. In return, the concessionaire pays the government, not the individual park or the NPS, a franchise fee subject to reconsideration every five years. At least theoretically, the agency uses industry figures to calculate percentages—for example Dunn and Bradstreet averages on profits for comparable services—but in reality field managers "negotiate nearly all the terms of concession agreements."[51] The fact that revenue does not go back into the park is important. Concessionaires may negotiate lower fees in exchange for underwriting capital improve-

FIGURE 2-2. *Appropriations following Fee Increases, Selected Fiscal Years, 1903–90*

Agency appropriations[a]

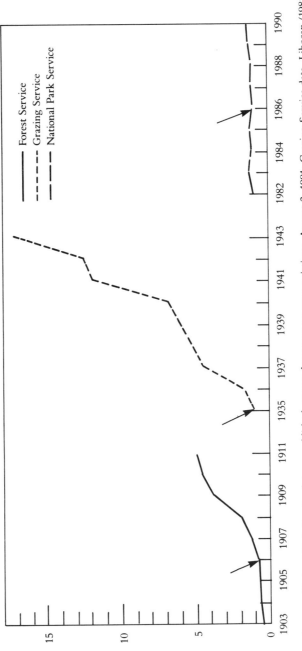

Sources: National Park Service data, unpublished agency document on appropriations, August 2, 1991; Grazing Service data, Libecap (1981b, p. 59); Forest Service data, Libecap (1981a, p. 158).

a. The scale of agency appropriations is based on appropriations in the year of fee increase as represented by 1. The arrows point to the fiscal year in which the fee increase occurred.

FIGURE 2-3. *Budget Outlays of Public Land Agencies, Fiscal Years 1981–88*

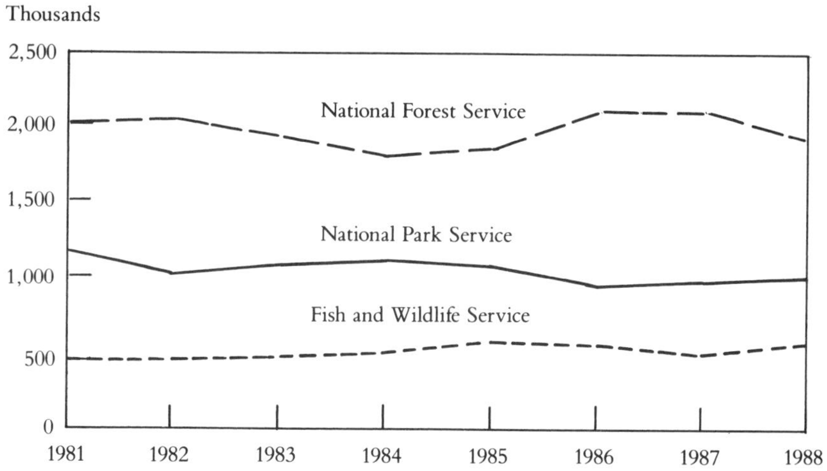

Thousands

National Forest Service

National Park Service

Fish and Wildlife Service

Source: *Budget of the United States Government*, various years, 1981–88. Outlay figures are taken from budget tables two years after the date shown to display actual money appropriated.

ments in the park. For example, a major concessionaire gets a relatively low franchise fee but is required to return 22 percent of gross revenues for park improvements.[52] The fact that managers, wanting to improve their parks but lacking enough funding, have historically negotiated low franchise fees is thus no surprise.

This bizarre relationship has been recognized for years and occasionally questioned. In 1976 a congressional study criticized the operations and contracts of concessionaires in national parks, but the report had little effect. The middle of the next decade also saw some controversy over these operations. Inspired by critical reports such as Bartlett's study of Yellowstone concessions, the trade association of National Park Concessioners fought back with its own reports. Controversy over concessions simmered until 1990.

The recent wave of attention to low concession fees was sparked by a perceived Japanese threat rather than by sudden awareness of the process. When Matsushita of Japan announced its intention to buy MCA Corporation in 1990, Interior secretary Lujan expressed dismay that one of MCA's subsidiaries was the concessionaire for Yosemite. Lujan used public concern over potential Japanese "ownership" of an American park to bring attention to the concessions contracts throughout the national park system.

The incident inspired several investigations of concessions contracts. Lujan's own task force recommended severe changes in concessions con-

TABLE 2-4. *Leading Concessions Operations in the NPS, 1992*
Millions of dollars unless otherwise specified

Concessionaires	Park unit	Gross receipts	Franchise fee[a]	Franchise fee[b]
Yosemite Park & Curry	Yosemite	83.5	0.62	0.75
AMFAC Hotels, Inc.	Grand Canyon	55.3	1.37	2.70
Circle Line Ferry	Statue of Liberty	13.8	1.38	10.00
Guest Services, Inc.	Sequoia	10.9	0.08	0.75
ARA Virginia	Shenandoah	9.2	0.26	3.00
Glacier Park	Glacier	7.5	0.11	1.50

Source: U.S. General Accounting office (1992c).
a. Nominal.
b. Percent.

tracts to make them shorter in duration and more competitive. In 1991 the GAO reported that park concessions in 1989 generated revenues of $531.5 million, but fees of only $11.5 million, a return of just 2.16 percent. Another report in early 1992 showed the seventy-five largest park concessionaires had 1990 sales of $486 million and fees of only $12.3 million, a 2.5 percent return.[53] A recent GAO investigation cited numerous sensational instances in which park concessionaires paid little or no fee for the use of government facilities. For example, Mount Vernon Inn on the George Washington Memorial Parkway grossed $4.3 million in 1990 but paid less than $115 a month in rent. Representative Mike Synar, Democrat of Oklahoma, commented: "These are amazing figures. . . . People dream about these kinds of opportunities."[54] Table 2-4 shows the largest concessions operations in the NPS and their rates of return to the federal government. The salient point is that increased business for these concessionaires did not translate to increased financial resources for the NPS.

Recent resolution of the Yosemite controversy illustrates two crucial facts regarding concessions contracts and the NPS. First, the notoriously small amounts of revenue returned to the government can be increased, but not systematically. The Yosemite contract was transferred temporarily to the nonprofit National Park Foundation in 1991. In late 1992 the contract was awarded to sports concessioner Delaware North Companies for roughly $60 million. Delaware North agreed to pay as much as 20 percent of gross receipts back into capital improvements for the park. This represents a substantial increase over the return to the U.S. Treasury from the previous concessioner, Curry Co.[55]

The Yosemite situation was an exceptional case, which involved considerable publicity. Systematic change will be much more difficult. According to two recent GAO investigations, NPS monitoring and oversight of concessions arrangements continues to be ineffective.[56] The agency lacks the capability to manage systematically, let alone change, the current concessions situation.

Second, even when individual contracts are changed, the NPS is not guaranteed any increased revenue. With the Yosemite contract and with others, capital improvements are in the eye of the beholder. For example, one Yellowstone concessionaire who was supposedly spending 22 percent of returns on maintaining government-owned buildings actually spent considerable amounts of the money on vending machines and snowmobiles.[57] Yosemite itself does not reap a financial reward from the new contract with Delaware North. Senator Dale Bumpers, Democrat of Arkansas, recently proposed the Concessions Policy Reform Act to mandate, among other things, the return of fee revenues to the parks. Without such legislation, the agency cannot expect a financial windfall from any renegotiated contracts.

Changes in the Operating Environment

When Mike Finley, now Superintendent of Yosemite, was managing the Everglades, he prepared congressional testimony opposing offshore oil drilling that might have harmed some of the endangered species within the park's boundaries. Before leaving for the oversight hearing, he received a phone call from the office of the secretary of the interior. The message was "Don't testify." When Finley told me the story, I asked him if it was an unusual case. He answered resignedly, "Nowadays it happens all the time."

Changes in the political environment affecting the NPS have altered the agency's operating environment. Conflicting goals and diminished political support have made the agency vulnerable to external influence, susceptible to second-guessing, and ultimately responsive in its behavior. The Finley anecdote suggests three important operational changes. First, high-level political actors have increasingly intervened in agency decisions. Second, the agency has responded, shifting its structure to accommodate these political demands. Third, employees like Finley are not particularly happy with the situation.

Under Political Assault

Political authorities in the legislative and executive branches possess a variety of tools that can be used to influence bureaucratic behavior. Many of these tools were applied with greater frequency to the NPS during the 1980s. As former director George Hartzog wrote in 1988, "Even as it struggles against the mounting threats to the integrity of our national

TABLE 2-5. *Congressional Hearings on the NPS, 1955–90*

Congress	Years	House	Senate	Total[a]
84th	1955–56	4	5	9
85th	1957–58	4	4	8
86th	1959–60	3	2	5
87th	1961–62	5	4	9
88th	1963–64	3	4	7
89th	1965–66	4	3	7
90th	1967–68	7	2	9
91st	1969–70	6	3	9
92d	1971–72	2	4	6
93d	1973–74	7	2	9
94th	1975–76	7	10	17
95th	1977–78	10	5	15
96th	1979–80	5	4	9
97th	1981–82	9	13	22
98th	1983–84	13	10	23
99th	1985–86	10	14	24
100th	1987–88	11	11	22
101st	1989–90	11	21	32

Sources: *CIS Annual* (Index) for each year; *CIS Annual* (Abstracts) for each year; *U.S. Serial Set Index* for 1959–69; and *U.S. Serial Set Index* for 1947–58.

a. Totals do not include appropriations hearings.

parks, the National Park Service itself has come under political assault."[58] When I asked Michael Frome, one of the foremost authorities on the NPS, if he thought that political involvement in the NPS had increased in recent years, he responded: "It's raw; now it's raw." Political tools of intervention include congressional oversight and case work (personal involvement), presidential control of personnel changes, and financial manipulations by both the legislative and the executive branches.

OVERSIGHT.   Congressional oversight of NPS behavior increased dramatically in the 1980s. Table 2-5 shows the number of oversight hearings for each Congress since the eighty-fourth began in 1955. These totals exclude routine hearings on appropriations, park expansions, and establishment of new parks. They include oversight of management of the NPS in general and of specific parks individually. Increased congressional involvement in park management is particularly apparent in the totals for the Congresses of the 1980s. The table cannot show the increasingly specific nature of many of these hearings. Whereas the hearings of the 1970s usually concerned broad issues such as the bicentennial events of 1976, hearings in the 1980s often focused on specific matters in individual parks such as management of bears in Glacier, geothermal development in Yellowstone, and waste sites in Utah.

As a part of its oversight function, Congress monitors agency behavior. Members often use outside sources such as the GAO to track and examine agency actions. One indicator of congressionally induced monitoring of NPS efforts is provided in table 2-6. As the investigative arm of Congress, the GAO formally reviews and evaluates public programs on request by members and committees. The table displays, both in simple numbers and in the increasingly critical nature of the titles, the expanded use of this particular tool in congressional oversight of NPS behavior in the 1980s. The increased number of GAO documents reflects greater GAO activity in general. Several of the GAO investigations were requested by members who were sympathetic to NPS problems. The increases reflect a greater proclivity on the part of members of Congress to monitor and ultimately to intervene in agency affairs.

CASE WORK.   Members of Congress can also influence agency behavior through case work, by getting personally involved in specific situations and decisions.[59] Case work is by its very nature anecdotal. Larry Henderson has worked for the NPS for thirty years. Now superintendent of Guadalupe Mountains, Henderson has served under seven presidents and eight NPS directors. He noted that in the past decade he had "seen more and more interest by Congress and Washington in NPS goings-on." Nearly all other interviewed employees of the NPS made comments similar to Henderson's observation. Many used the term micromanagement to describe cases of political involvement down to the smallest details. Zion Superintendent Wiese said explicitly that his "autonomy was becoming less and less because of micromanagement by Congress and the regional office." Most of the people interviewed said that the increases had occurred within the past ten years.

While prominent cases have been covered by the media, less visible and fairly simple incidents illustrate micromanagement. When I visited Grand Canyon, Ken Miller, the chief ranger, described a situation facing him. A teacher wanted to bring an ecology class with her on a visit. The class exceeded the group size (forty-eight) specified in the park's regulations. She had solicited congressional help to seek an exemption. Miller anticipated having to write four letters to individual members (two senators, the representative from her district, and the representative from the Grand Canyon district) to justify NPS action. Another example: Saguaro superintendent Bill Paleck was recently under pressure about a proposed route for traffic to downtown Tucson. Paleck urged the town to build the route around the park. Commuters, seeking a shortcut through parkland,

TABLE 2-6. *GAO Reports on the National Park Service, 1970–90*[1]

Year	Title
1970	"Problems in Land Acquisitions for National Recreation Areas"
1971	"Problems Associated with Development of Burns Waterway"
1975	"Operations of the John F. Kennedy Center for the Performing Arts" "Concessions Operations in the National Parks: Improvements Needed in Administration"
1979	"Why the National Park Service's Appropriations Request Process Makes Congressional Oversight Difficult"
1980	"Better Management of National Park Concessions Can Improve Services Provided to the Public" "Review of Allegations of Preferential Treatment"
1981	"Establishing Development Ceilings for all National Park Service Units" "The National Park Service Should Improve Its Land Acquisition and Management at the Fire Island National Seashore" "Weaknesses in Internal Financial and Accounting" "Impact of Proposed Impoundment of Funds" "Mining on National Park Service Lands: What Is at Stake?"
1982	"Increasing Entrance Fees" "Interior's Program to Review Withdrawn Federal Lands: Limited Progress and Results"
1983	"National Parks' Health and Safety Problems Given Priority: Cost Estimates and Safety Management Could Be Improved" "National Park Service Has Improved Facilities at Twelve Park Areas"
1984	"Additional Actions Taken to Control Marijuana Cultivation and Other Crimes on Federal Lands" "National Park Service Needs a Maintenance Management System"

complained to their congressional representatives, who then contacted Paleck.

How often these situations occur varies considerably. Superintendent Noel Poe of Arches estimated that he heard from a member of Congress about every other week on some matter. Others suggested that it happens even more frequently. Superintendent Rob Arnberger of Big Bend calls it "regular business."

Sources on Capitol Hill admit that congressional intervention in specific park affairs in recent years has been "impressive." They attribute increased involvement to several factors. First, many members do see the parks as a source of electoral opportunities, particularly in terms of economic development. Second, even sympathetic members accelerated their interactions with the agency during the 1980s out of protective concern for the hostility displayed by the Department of the Interior. Third, NPS

TABLE 2-6. *Continued*

Year	Title
1985	"New Rules for Protecting Land in the National Park System: Consistent Compliance Needed"
1986	"National Parks: Law Enforcement Capability and Cost Comparisons at Two Recreation Areas"
	"National Parks: Emergency Law Enforcement Expenditures at Two Recreation Areas"
	"Parks and Recreation: Obligations and Outlays from the Land and Water Conservation Fund"
	"Parks and Recreation: Recreational Fee Authorizations, Prohibitions, and Limitations"
	"Restoration of the Statue of Liberty National Monument"
	"Construction Contract at Jean Lafitte National Historical Park"
	"Concerns Raised about National Park Service Actions at Delaware Water Gap"
1987	"Limited Progress Made in Documenting and Mitigating Threats to the Parks"
	"Patowmack Canal Preservation Responsibilities Being Met by Park Service"
1988	"Park Service Managers Report Shortfalls in Maintenance Funding"
	"Parks and Recreation: Interior Did Not Comply with Legal Requirements"
	"Proposed Alaska Land Exchanges"
1990	"Federal Fire Management: Limited Progress in Restarting the Prescribed Fire Program"
	"Recreation Concessioners Operating on Federal Lands"
	"Improvements Needed in Managing Concessioners"
	"Cost Estimates for Two Proposed Facilities in Texas"

Source: U.S. General Accounting office monthly reports and indexes.
a. The GAO has little confidence in its database before 1967.

officials themselves are not without blame. Bereft of high-level support, individual managers and superintendents have cultivated personal relationships with individual members of Congress to maintain some influence on legislative or spending decisions.

PERSONNEL CONTROL.   Presidential administrations use personnel changes and procedures to alter bureaucratic behavior. Administrations have used appointments, censorship of behavior, and manipulation of career paths to influence the NPS.

The obvious means for personnel control is through the appointment process. Roughly 3 percent of federal employees are political appointees. Administrations can thus affect agencies by determining who gets placed in positions of power. President Richard Nixon broke tradition by appointing Ron Walker, who was not a veteran of the NPS, as its director in 1973.

Careerists perceived the appointment as a substitution of politics for experience. William Penn Mott was appointed in 1985 after experience as director of California state parks under Governor Ronald Reagan. President George Bush repeated the mistake of appointing an outsider when he named James Ridenour in 1989. Ridenour's key qualification for the job was that he had been a campaign fund-raiser for Vice President Dan Quayle in Indiana. The appointment of outsiders rather than careerists to head the agency may well enhance political ties rather than organizational loyalties. According to Sandy Rives, a former Washington official who is now assistant manager of Shenandoah, the continuing appointment of noncareerists logically "leads to more decisions being affected by politics," since these individuals come to the job with different incentives and career experiences. This recent practice of hiring outsiders continued when President Bill Clinton appointed Smithsonian official Roger Kennedy in 1993 after considering such other notables as Robert Redford and Tom Brokaw.

Second, top administration officials, particularly those in the DOI, have increasingly censored employee behavior in recent years. The story of Howard Chapman is illustrative. Chapman was Western Region director in the mid-1980s until he criticized enough Reagan administration actions to draw the wrath of William Horn, assistant secretary of the Department of the Interior. Horn downgraded Chapman's efficiency rating until Chapman threatened to sue. Chapman was then, in former director Hartzog's words, "harassed into retirement."[60] Such action was consistent with the Reagan administration's control of personnel behavior. Frome noted that the Reagan administration "did more to inhibit individuality and initiative on behalf of national parks preservation than any earlier administration."[61] Shenandoah's Rives, who was working in the D.C. offices in the early 1980s, offers inside confirmation of this view: "I was there then, and they just stomped on the parks."

A recent example shows that top-level censorship did not stop with the end of Reagan's second term. Lorraine Mintzmyer was a director of the Rocky Mountain Region in 1991, and she supervised preparation of a planning document for the future of Yellowstone, *Vision for the Greater Yellowstone Area*. The document called for greater resource protection through cooperative efforts of the NPS and the Forest Service on surrounding lands. This emphasis upset local private interests and inspired attacks from Senator Alan Simpson, Republican of Wyoming, and White House Chief of Staff John Sununu. Mintzmyer testified that a DOI appointee rewrote the document to be more politically sensitive and that the final product was reduced from sixty to ten pages. When Mintzmyer objected, she was transferred to Philadelphia. She eventually resigned. In fact, all principals involved in original preparation of the document were transferred or resigned. Secretary Lujan said there was "no connection."

A subsequent congressional investigation revealed DOI memos that confirmed the political censorship of the report. Mintzmyer now says, "This type of political intervention has led to such a dilution of the Park Service purpose that the service's survival is at stake."[62]

Third, rather than censoring individual personnel, administrations have recently attempted to control NPS personnel by altering career paths. The Reagan administration, according to some sources, attempted for years to manipulate the transfers and evaluations of the senior professional ranks of the agency. The attempted changes would have replaced dissenters with personnel holding similar opinions on policy issues. That philosophy continued even after Reagan left. A Bush administration proposal would have given DOI political appointees authority to appoint or transfer any NPS employees at General Schedule (GS) level fourteen. This would have included staffers at regional and central levels and most superintendents. The most recent individual situation involved the use of fifty-year old legislation that allows congressional employees to move into the executive branch without going through the regular hiring procedures. At the end of 1992, House staffer Stephen Hodapp sought and received the position of assistant superintendent at Shenandoah. Superintendent Wade explicitly stated in a letter that he did not want Hodapp but was ordered by DOI appointees to hire him anyway. Director Ridenour admitted that he "strongly encouraged" the agency to find a place for Hodapp.[63] Field personnel despise such manipulations. As Zion's Wiese said, "Many quality people in the field are not getting a chance because of too many political appointees."

FINANCIAL CONTROL.   Agency budgets in recent years reflect greater input from members of the legislative and executive branches. Even high-ranking officials in budgeting admit that their process is more affected than it used to be. When I talked to Comptroller General Bruce Sheaffer, he spoke so highly of his budget procedures that I asked him if budgets were somehow more insulated than other aspects of NPS operations. He responded: "Budgets are not insulated from politics. In fact, political influence is probably increasing in budgets." B. J. Griffin, director of operations in the Western Region, corroborated that viewpoint from the regional level. "Members have a lot more to say now on the complexion of the Park Service budget."

Two recent procedural developments have stimulated greater political intervention in NPS budgets. The first of these was, perhaps ironically, the increased revenue resulting from higher entrance fees. This result is actually not that surprising given the experiences of other public lands agencies in similar circumstances. At various times earlier in the century, both the Forest Service and the old Grazing Service experienced increased political involvement in agency decisions, along with their increased appropriations, as the result of increased user fees. NPS fee increases did not result in increased appropriations, but they did stimulate congressional interest in expenditures. With appropriation levels flat, increased congressional manipulation of existing amounts translates to less discretionary spending available to the agency.

Evidence that fee increases contributed to congressional intervention in NPS behavior is limited for several reasons. First, intervention occurs on a specific, case-by-case basis and is thus not systematic. Second, the changes are quite recent, and the effect of political involvement may take years to manifest itself fully, as it did in the case of both the Forest Service and the Grazing Service. Third, congressional involvement in NPS behavior was increasing even without the entrance fee controversy, and the contribution of fee increases to this phenomenon is difficult to consider separately. Nevertheless, the evidence that is available supports the hypothesized relationship.

The most direct evidence for the asserted relationship between fees and political intervention is the observations of NPS field managers. Superintendent Wickware of Canyonlands put it most explicitly: "The fee system is a charade. It's here to stay because it's a windfall for Congress, but it's a negative for the National Park Service." Several other superintendents told me of their negative feelings about the political attention that accompanied fee increases. One superintendent explained the exact process this way. Rather than returning the fee money to the parks for discretionary operations, members of Congress instead observe the increased revenues in the General Fund and then, under pressure from various constituents, tell field managers what to spend the money on within individual park units. Thus even though appropriations did not go up accordingly, more expenditures were designated for specific projects, such as visitor centers or roads, from higher levels. Increased fees thus provide electoral opportunities. Another official said that he and other superintendents had argued against the fee increases, though they had not done so in formal testimony, precisely because they knew the higher revenues would encourage increased micromanagement by members of Congress.

Documentary evidence supports these views. Budgets before the entrance fee increases are fairly explicit in the sections on Constructions and Land Acquisitions, but they rarely make specific designations in project planning. In 1986 and 1987 appropriations committees specified certain projects, particularly visitor centers and campgrounds in individual parks. By 1988 planning constituted a whole new section, with more than two dozen projects involving nearly $16,000,000 designated explicitly for items such as visitor centers at Big Thicket and Florissant Fossil Beds and a campground at Indiana Dunes.[64]

Even the official line regarding fee increases allows that the issue may have stimulated congressional intervention. Though the director of the budget for the NPS told me that the superintendents were probably overstating their case, he did allow that such intervention would not happen through his office but rather on an individual, informal basis. He even suggested the procedure whereby a member would contact the superintendent, remind him of his support for higher fees, and then propose how that money be spent. As in similar cases in the past, the fees provide an excuse for members to act as ombudsmen. As one superintendent told me, constituents are more likely since the fee increases to complain to their representatives about their park experience since they assume that the money they paid to enter the park goes back into the park's budget.

Political intervention in agency finances is also stimulated by concessions contracts. Relationships between politicians and concessioners have been characterized in the past as "notorious." The political clout of concessioners in the 1980s was hardly diminished. At a business conference in 1981, Secretary Watt heard a complaint about a park official from a concessionaire and responded: "If a personality is giving you a problem, we're going to get rid of the problem or the personality, whichever is faster." Even at the end of the decade, Interior's own task force on concessions advised that "poor relations with concessioners are generally not viewed by park personnel as career-enhancing." Field-level employees say that concessioners will go to the NPS director for help and then to Congress if need be to resolve a problem.[65] Some are quite well connected. Allan Howe, for example, the Washington manager of the Conference on National Park Concessioners, is a former U.S. representative from Utah.

The kind of political intervention that will be inspired by the recent controversies over concessions contracts remains to be seen. The many different sources demanding greater coordination of concession arrangements were at least partially rewarded in the Yosemite case. Even with the new Yosemite contract, the new concessioner retains the preferential right of renewal when the contract expires. This "future lack of competition" left the new concessioner with more freedom in the park than some observers would have liked. Agency influence over concessioner operations is also likely to be diminished by the fact that the negotiations were handled by Secretary Lujan, rather than by the NPS. Increased central involvement in negotiating contracts may mean that field managers will lose the ability to negotiate compensation arrangements other than fees, arrangements such as capital improvements, which they used in the past.[66]

Secretary Lujan initiated the process to revise contracting procedures shortly before the end of the Bush administration. Lujan's announced changes, published in the *Federal Register* in 1991, called for ending preferential renewal of contract leases and for revising the possessory interest clause of the 1965 law. These proposals immediately inspired criticism from the concessioners themselves and from members of Congress. Concessioner official Howe called his organization "very disappointed." For his part, Lujan made his own intentions clear: "We're not looking to break anyone or be unreasonable. I am very happy with the system. I just want a greater return."[67] Obviously Lujan was not interested in reducing the ties between concessioners and political authorities or in increasing the amount of discretionary revenue returned directly to the parks.

Eating Trail Dust

The internal structure of the NPS has changed in response to the political assault of the past fifteen years. The decentralized structure of the 1970s has evolved into an organization in which decisions are increasingly made at top levels of the agency or by political authorities outside the NPS. Saguaro Superintendent Paleck says the analogy of the superintendent as ship captain should be replaced by that of a trail boss on a cattle drive. As Paleck says of his job, "Sometimes you have to eat trail dust." Or, to interject a historical note into Paleck's analogy, Senator Hansen's cows have moved to the front.

AGENCY STRUCTURAL RESPONSES.   In many ways, the structural responses of the NPS to increased political intervention were quite logical. Frequent appearances by agency officials before congressional committees create centralizing pressures in that the officials are expected to provide information and answers that are facilitated by data collection at central offices. Furthermore, executive branch officials may fear embarrassment resulting from being held responsible for decisions made by subordinates and will thus take more interest in field decisions. More frequent monitoring can stimulate centralization through increases in office staff for public relations and more required clearances for field actions. The GAO often recommends agency action that requires central coordination. For example, in a 1987 investigation, the GAO stressed the need for "the Park Service to develop a comprehensive, system-wide approach to protect and manage park resources and provide the basis to make more informed funding decisions."[68] Casework and manipulation of personnel and finances stimulates agency awareness of high-level preferences for uses of labor and capital.

Actual structural changes in the NPS are not immediately apparent in formal organization charts. The current formal structure of the NPS is displayed in figure 2-4 and figure 2-5. As figure 2-4 shows, the agency is subdivided into ten regional offices, which have headquarters in Anchorage, Seattle, San Francisco, Denver, Santa Fe, Omaha, Atlanta, Philadel-

FIGURE 2-4. *Formal Structure of the NPS, 1991*

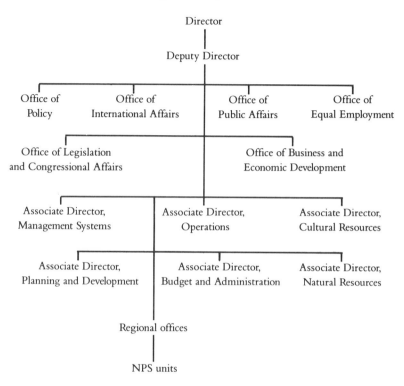

Source: Unpublished U.S. NPS document, 1991.

phia, Washington, D.C., and Boston. The organization chart for the Western Region is pictured in figure 2-5 to illustrate a typical regional structure. Regional offices "serve as the primary communications link between the [Washington] Office of the Director and park areas."[69]

Though the formal organization chart for the agency does not appear dramatically different from that of the 1970s (figure 2-1), some subtle differences are present. The Office of Legislation is now the Office of Legislation and Congressional Affairs. The agency now has a Tourism Division as well as a Visitor Services Division. The Office of Planning is now part of Planning and Development, headed by an associate director, rather than part of the director's central staff. The Office of Cooperative Activities has been replaced by the Office of Business and Economic Development. New is the Office of International Affairs. These changes reflect political realities, but they also show the importance of certain economic concerns that were at the heart of the philosophies of the Reagan and Bush administrations. Political interests have altered the structure of the NPS.

FIGURE 2-5. *Formal Structure of an NPS Regional Office, 1990*

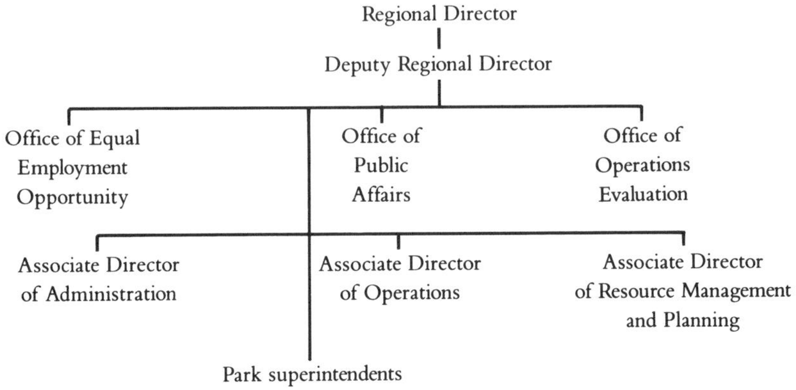

Regional Director

|

Deputy Regional Director

Office of Equal Employment Opportunity	Office of Public Affairs	Office of Operations Evaluation
Associate Director of Administration	Associate Director of Operations	Associate Director of Resource Management and Planning

Park superintendents

Source: U.S. National Park Service (1990c, p. 1).

One inevitable result of the structural changes depicted in figure 2-4 that is not readily apparent in the chart but that contributes to agency centralization is the effect on allocation of personnel. The increase in the number of headquarters offices and divisions translates to an increase in personnel in Washington. As the agency's own Vail Agenda report admits, "Today, nearly one-third of all employees (including all twenty-one senior executives) work outside the structure of the parks units." This has occurred in spite of the fact that, as the report states, "virtually all of its [NPS] value-creating activities are centered in the individual park units and programs."[70]

Examining the operating budgets for the NPS over time also provides some revealing evidence. Between 1980 and 1988, the field portion of the operating budget, that going to the parks, declined from 68 percent to 51 percent. During the same time period, the portion allotted to general administration increased from 9.1 percent to 10.3 percent. This included increases of more than 100 percent in public affairs, roughly 100 percent in expenses such as telephones and postage, and 77 percent in central office administration. The largest growth of any activity in the operations budget occurred in the nonpark portion of park management, which increased from 21 percent to 32 percent. This latter category includes such items as regional maintenance programs and centralized drug control initiatives. Centralization of these expenditures obviously means less discretionary spending at the field level. As the Vail Agenda states, "[NPS] management budgets are channeled into special projects and excessive administrative layers divorced from core functions of the park units and other agency programs."[71]

EMPLOYEE APPRAISALS.   Testimony for the changes in agency structure is provided by the comments of employees. Thirteen years after I had met the superintendent of Yosemite, Les Arnberger, I met his son, Rob, the superintendent of Big Bend. When I found him, he was making popcorn for the staff. I had heard a lot about the power of superintendents within their own jurisdictions, and since this was the first time I had interviewed one, I was somewhat surprised to find him in the kitchen. Arnberger smiled when I asked him about that. "Much of the time I can do what I want around here, but on Friday afternoons," he rattled the popcorn maker, "they expect me to do this." Symbolically at least, the contrast between the two meetings I had with the members of the Arnberger family suggests that demands on national park managers have changed.

Nearly 90 percent of the NPS field-level employees I interviewed told me that their actions and decisions have been increasingly affected in recent years by politicians, headquarters, and even regional offices. This view was cited at the parks that are the crown jewels, such as Yosemite, as well as at relatively obscure national monuments. As one manager told me, the decisionmaking process now involves "politics within politics within politics." When I asked one field employee if he agreed with the central office assessment that 80 percent of daily decisions were made without Washington's input, he said, "Maybe, but that other 20 percent is absolutely critical." Finally, one trail-level bureaucrat told me that the real effects of centralization would be experienced in the future: central pressures were increasing and, within the next decade, those people in the parks "who know how to deal with it" will have left or retired.

Officials in several regions were interviewed. At the regional level, personnel were, not surprisingly, more guarded in their comments than their field employees, but they did admit that there is serious potential for centralization pressures from their own regional offices. For example, all project design requests from the parks, such as those for new facilities, must be sent to the regional office. If the project costs more than $250,000, the region sends it along to higher levels. Otherwise the regional office becomes heavily involved in funding, planning, and coordinating the project. Further, regional offices are closer to high political levels in more ways than just chain of command. The Rocky Mountain Region's deputy director, for example, cited growing frustration at having to answer to increasingly partisan political appointees. Others agreed with the general assessment of increased centralization but cited wide variations among regions.

Reportedly, officials in some regions are attempting to reverse some of the centralization trends of recent years. As yet, their efforts have been neither noticed nor appreciated at the field level. In fact, field employees are quite opinionated about regional office intervention. One park manager who has worked at all three levels of the organization understandably wanted to remain anonymous when he described the continual intervention by regional officials in field decisions: "They do it all the time; they have to justify their existence."

EXTERNAL APPRAISALS.   Though external appraisals of recent NPS behavior concentrate more on the cause (political pressure) than on the result (structural changes), they support the argument that the agency has responded structurally to political demands. "The Park Service is not an autonomously expert agency but more of a responsive one. . . . The parks mean too many things to too many different groups and individuals to expect that this should be otherwise."[72] An independent review of nine prominent works on the NPS concludes that the agency has indeed become significantly more "responsive to politics."[73] In becoming more responsive, the structure and the operations of the agency have shifted dramatically. The experience of the NPS thus illustrates an important conclusion of organization theory literature: "The political environment within which governing takes place is normally regarded as the most centralizing determinant of public administration."[74]

Long-term observers of the park system criticize the structural evolution for enhancing the responsiveness of the agency to central concerns but diminishing field-level discretion. "Congressional initiatives, even much needed ones, coupled with the constraints of executive policy, have sometimes hemmed in the service's discretion and diminished its opportunities to exercise leadership and flexibility."[75] The most authoritative comparison of world park systems summarized recent changes this way: "The NPS established in its early years a reputation powered by professional ethics, largely free of political considerations. Recent times have seen personnel selection, designation of new parks, and management practices in parks more seriously influenced by special-interest politics."[76] A critic who described the historical autonomy of Yellowstone now writes that "influence over policy has fallen by default into the hands of a small coterie of featherbedding Congressmen, bigwig concessionaires, and grey-flanneled Beltway conservationists."[77]

Finally, other political actors bemoan the fact that while the NPS has centralized in response to political pressures, it has not done so efficaciously. One congressional source described the NPS as "notoriously unsophisticated" in its dealings with members and committees. The NPS has considerably less experience in educating and training its personnel for congressional interaction than do other agencies, such as the Forest Service. In a self-portrait of the agency, employees of the NPS admit: "At the level of the overall system, the Park Service is variously seen as run and overrun by Congress, the White House, the Secretary of the Interior, private interest groups, or public interest groups."[78]

An All-Time Low

The NPS field manager will remain anonymous as he wished. Even though he was a veteran of more than two decades with the agency, his statements were angry, often bitter. He spoke of what he called the "dark side" of the NPS, and he backed up his views with off-the-record specifics. I finally commented that the agency did not sound like the one I had known as an employee fifteen years before. "It's not just me," he said coldly, "Morale is at an all-time low."

The NPS has always been famous for the commitment of its employees. As Bernard Devoto wrote in 1953: "The most valuable asset the Service has ever had is the morale of its employees." Thirty years later, a survey of public land agencies concluded: "The strength of the [NPS] organization undoubtedly lies in the commitment of its personnel."[79] Today, while commitment remains intact, the morale of those same employees is, in the view of many park employees, poor. The reasons for low morale, closely interrelated with the recent changes in the agency, include turnover of leadership, ideological frustration, and low compensation.

LEADERSHIP TURNOVER.   Increased turnover at the top has at least two effects. First, subordinates may become frustrated having to deal with changing directives and priorities. Second, high turnover may lead directly to reduction of responsibilities among field employees. As new executives try to gain control of agency actions, they impose commands and directives that influence employee behavior. These actions provide a centralizing pressure, and, by taking discretionary authority away from subordinates, may foster resentment.

High turnover among top officials in recent years is easily documented. As table 2-7 shows, in its first fifty-five years, the NPS had seven different directors. Within the past twenty years, the agency has also had seven directors. Only in the decades of the 1970s and 1980s did the NPS undergo three transitions at the top level. Furthermore, since the NPS is an agency within the Department of the Interior, an examination of turnover at high levels must also consider changes at the top of this cabinet department. Interior was headed by conservation-oriented Cecil Andrus at the start of the decade, then by Watt, and then by three other secretaries in the 1980s, making a total of five for the decade. Interior also had five secretaries in the 1970s. The 1920s were the only other decade in which this happened.

TABLE 2-7. *Directors of the National Park Service*

Director	Tenure
Stephen T. Mather	5/16/17 – 1/8/29
Horace M. Albright	1/12/29 – 8/9/33
Arno B. Cammerer	8/10/33 – 8/9/40
Newton B. Drury	8/20/40 – 3/31/51
Arthur E. Demaray	4/1/51 – 12/8/51
Conrad L. Wirth	12/9/51 – 1/7/64
George B. Hartzog	1/8/64 – 12/31/72
Ronald H. Walker	1/7/73 – 1/3/75
Gary E. Everhardt	1/13/75 – 5/27/77
William J. Whalen	7/5/77 – 5/13/80
Russell E. Dickenson	5/15/80 – 3/3/85
William P. Mott, Jr.	5/17/85 – 4/16/89
James M. Ridenour	4/17/89 – 1/20/93
Roger Kennedy	6/93 – present

Sources: Sontag (1990, p. 64); U.S. National Park Service, Office of Public Affairs.

Many field employees think the high turnover occurs because political appointees are simply padding their credentials. They point to the influx and short tenure of recent outsiders in the top positions of the agency. This impression is consistent with Hugh Heclo's comment years ago: "The single most obvious characteristic of Washington's political appointees is their transience." A former NPS director describes the effect of the uncertainties that come with high turnover at the top levels: "employees are dispirited, lacking in confidence, and marking time."[80]

IDEOLOGICAL FRUSTRATION.   A second reason for the drop in morale is ideological. A 1980 survey showed most agency personnel favoring preservation as "the major purpose" of the NPS. Obviously these employees were at least uncomfortable with the attitudes of their bosses in the Department of the Interior a year later. One superintendent told me: "Watt set us back thirty years." Other field personnel had running battles with high-ranking officials. The dispute between forty-year veteran Howard Chapman and DOI assistant secretary Horn was particularly contentious. Another superintendent said of President Bush, "He says he's an environmentalist, but he doesn't even know what it means." During the 1980s careerists in the NPS fought constantly with their supervisors. As NPS director Russell Dickenson, a thirty-year veteran, later recalled, "Every operating principle established throughout our working lifetimes was challenged."[81]

Most NPS employees recognize the growing threats facing the parks but bemoan the inadequacy of resources with which to meet them. The funds for park management have not even kept pace with inflation. As the director of the NPS budget admits, the parks must now compete for the limited funds that are available. This kind of competition does anything but enhance the esprit de corps of the agency.

INADEQUATE COMPENSATION.   A third reason for low morale is material reward. NPS employees recognize the perquisites of their job locations, but that compensation may not be enough to maintain high morale. As Teton superintendent Jack Neckels said, "I can't pay my kid's tuition on Teton sunsets." Actual pay is low. The vast majority of NPS personnel are on the GS pay scale in various brackets. The average GS salary in 1991 for all NPS employees was just over $27,000, but field rangers make much less. In fact, half of the full-time rangers were classified at the GS-7 level, with an annual salary of $21,906.[82] Many rangers with families supplement this low income with other jobs or with food stamps. Housing is available, but often in poor condition; the NPS estimates that more than $500 million is needed to recondition employee housing. Finally, with cuts in the work force, most employees see few opportunities for advancement. As one superintendent told me, many good people don't get promoted. Even while commitment among most remains high, the inevitable result is higher turnover. Annual turnover rates among NPS employees rose from 3 percent in the 1970s to 8 percent in the early 1990s.

The drop in morale has elicited scathing indictments from employees themselves. Recent published accounts emanate from former rangers such as Runte and Shanks, former superintendents such as Jerry Schober of Gateway Arch, regional directors Chapman and Mintzmyer, and even former NPS directors. When I asked Dan Cockrum, maintenance manager for Grand Canyon and a twenty-eight-year veteran, about recent changes in the NPS, he responded: "We're not the family we used to be." In the 1992 Vail Agenda, other observers used the same term. "All these factors and others have eroded the sense of 'family' that was once at the heart of the Park Service culture." The result, according to the document, of inadequate training, thwarted initiatives, and mismatched demands and resources is that "the National Park Service faces significant morale and performance problems."[83]

Changes in Policy Behavior

How have these recent changes in the NPS affected policy behavior? The theoretical argument proposed in chapter 1 suggests that the NPS, operating in an environment of severe political pressure, structural centralization, and low morale, will make decisions and take actions based on short-term concerns with political fallout. In effect, political authorities will use the parks for electoral and economic purposes rather than allowing the NPS to preserve natural conditions. These changes are evident in system expansion, management policies, and the allocation of resources.

Lost Control

At Steamtown National Historic Site, more hard hats are apparent than ranger hats. Construction hard hats are necessary because this "historical" site is literally being constructed out of a train yard in Scranton, Pennsylvania. The site and the funds for its construction are the result of political clout. The controversial Steamtown is a graphic example of the process for adding new park units to the American system.

The obvious place to begin examining NPS behavior is the creation of new parks. After all, parks must exist before questions concerning managing and funding are even possible. How are emphases on preservation or use manifest in the creation of new parks? Preservation is evident when the system is expanded to preserve diverse ecosystems. "The central task of the NPS is to maintain, within a world that is mostly governed by human activity, a substantial sample of functioning natural systems that are not characterized by human domination and to protect major examples of past human activity so that present and future generations can learn what the past has to offer them."[84] Emphasis on use is apparent, however, when new parks are established because creation is useful to people, particularly politicians.

CRITERIA.   What are the criteria for becoming an NPS unit? The most basic factor is that the place must be unique in some way. The NPS operationalizes this characteristic by requiring that each unit "represent some significant aspect of our natural or cultural heritage or an outstanding recreational resource."[85] Establishing a more precise definition that is systematic and consistent is difficult because the NPS manages a variety of places and because the most important factor in new park creations in recent years has been not uniqueness or national significance but political and economic clout.

The NPS manages a variety of units. Table 2-8 lists the number of units within each of twenty different categories and the total acreage for each category. Many of these units, including all the national parks, were established by Congress through individual pieces of legislation. National monuments are created by the president under the authority of the Antiquities Act of 1906. National historic sites are designated by the secretary of the interior. The category of protection is important. For example,

TABLE 2-8. *Summary of Units of the NPS, 1990*

Type	Number	Acreage
International historic site	1	35
National battlefields	11	12,772
National battlefield parks	3	8,767
National battlefield sites	1	1
National capital park	1	6,469
National historic sites	69	18,468
National historical parks	29	151,633
National lakeshores	4	227,244
National mall	1	146
National memorials	25	7,949
National military parks	9	34,047
National monuments	78	4,844,610
National parks	50	47,319,321
National parkways	4	168,618
National preserves	13	22,155,498
National recreation areas	18	3,686,923
National rivers	5	360,630
National scenic trails	3	172,203
National seashores	10	597,096
National wild and scenic rivers	10	292,597
Others	11	40,138
Total[a]	358	79,997,167

Sources: U.S. National Park Service (1991c, p. 1) and (1989b, p. 13).

a. Categories may overlap.

national preserves like Big Cypress are lands bordering national parks (Everglades) and allowing more recreational uses than the parks.

On paper, the process of creating new units is as follows. An individual, group, or even institution can propose an area for inclusion in the national park system. The NPS, particularly the Division of Park Planning and Protection, is then responsible for determining the qualifications of that proposal. The NPS uses the following criteria for its evaluation. To be nationally significant, a proposed area should exemplify a particular type of resource, possess exceptional heritage value, provide superlative recreational opportunities, and be relatively unspoiled. To be suitable, an area must represent a type not already protected. To be feasible, an area "must have potential for efficient administration at a reasonable cost."[86] The NPS, using normally a two-year review, determines the proposal's merit and then considers other management options such as designation as a state park. The NPS study of the proposed area is then used to determine (by Congress for parks, by the DOI for historical sites) whether the unit will be included in the system and to establish specifics such as boundaries.

In the early 1970s environmentalists emphasized the preservation of representative areas. One urged "a thematic approach to the identification and assessment of natural areas." The NPS responded with a thorough study of the system's composition. The agency concluded in the *National Park System Plan* of 1972 that many historical themes and types of natural regions such as plains lacked representation in the current system. The study recommended adding more than 200 additional units.[87]

REALITY.   In practice the process is heavily affected by the political system. New units are added or passed over according to political determination of usefulness, including benefits at the polls, rather than scientific emphasis on preservation of representative areas. To be accurate, political effect on unit creation is not a new phenomenon. What has been different in the past fifteen years is the brazenness of political calculations in the face of other proposed procedures for adding new units to the system.

The importance of political calculations was openly displayed at the end of the 1970s in two separate pieces of legislation. The first was almost humorous. The 1978 National Parks and Recreation Act became known as the Park Barrel bill "because it provided for so many new projects in so many states." In debate on the floor, Sen. Robert Dole, Republican of Kansas, asked: "Is there any state other than Kansas that did not end up with a park?" The floor manager, James Abourezk, Democrat of South Dakota) answered, "Did we leave you out, Bob?" Dole countered, "I have two more years in my term." Amidst the laughter, Senator Hansen, Republican of Wyoming, offered: "It is my understanding that six states— five others—did not make it." Dole responded, "I appreciate that. We will have a meeting later."[88] Even Guam's representative would not have been invited to Dole's meeting because the War in the Pacific National Historical Park was created by the Park Barrel.

The second massive creation of new parks at the beginning of the 1980s was more contentious than the Park Barrel but just as affected by political maneuvering. The creation of new park units in Alaska stimulated one of the most bitter disputes over public lands ever witnessed in the United States. Various lands had been protected until 1978 by the Alaska Native Claims Settlement Act. Before that act expired, environmentalists and their allies in Congress tried to preserve more than 100 million acres from developers and resource extractors by designating parcels of land as NPS units. The negative response from economic interests and from Alaska's own senators was intense. After proposed legislation was stopped by a filibuster in 1978, President Carter used the authority granted him by the Antiquities Act to create seventeen national monuments and twelve wildlife refuges. Since this protection is not as secure as national park status, proponents continued to try to secure permanent legislation. The bitter fight lasted until the elections of 1980. Then, with the pro-development Reagan administration on its way to Washington, preservationists were forced to accept a compromise version of the Alaska National Interest Lands Conservation Act. The fact that the bill became law on December 2, one month after the elections and one month before Reagan's inauguration, is not a coincidence. As Representative Morris Udall, Democrat of Arizona, one of the sponsors, said at a news conference following House passage, "I don't claim that this is a great victory . . . it is just political reality to act now."[89] The bill created 44 million acres of new parks and 54 million acres of new wildlife refuges. However, many of the acres were mandated to allow hunting of wildlife. Other uses not traditionally encouraged in national parks, such as use by snowmobiles, were accepted with the compromise. Timber development and oil exploration were allowed on many of the lands. In the largest expansion ever of a park system, the creation process was ultimately determined by political reality.

The Alaska bill was as close as preservationists would get in the 1980s to pursuit of a representative park system. As described earlier, Secretary Watt and other political decisionmakers of the decade blocked numerous efforts to create new parks by simply refusing to spend the purchase money. In 1981 testimony before Congress, Watt said: "My definition of stewardship in the Reagan administration is that we must take care of what we have charge of before we take further lands off the tax rolls for purposes that will not be developed for a good number of years."[90] The Land and Water Conservation Fund had been established in 1965 to use money from offshore oil leases to purchase new parklands and refuges. Watt froze nearly one billion dollars of these funds, even though the NPS had already selected more than 400,000 acres of private land within parks for purchase. Among the parcels of land in limbo were thousands of acres within the Santa Monica Mountains Recreation Area. Even after Watt left, this acreage, now inflated in price, was not purchased. As a result, only 14,000 of the 150,000 acres within the area were federally owned by the end of the decade. Through the 1980s Congress appropriated, on average, only $200 million for land acquisition even though the fund was authorized at a rate of $900 million a year.

Many of the new park units that were created during the 1980s reflect the use orientation of politicians. Biscayne and Channel Islands, both designated in 1980, continue to allow boating and development mainly because of large nonfederal holdings in the parks. Nearly half of the acreage in Biscayne and almost three-fourths of Channel Islands are nonfederal lands and thus available to economic interests.[91] Even Great Basin, the only national park created during the Reagan administration, was established as a result of political maneuvers. The park, in a magnificent mountain section of eastern Nevada, had been proposed for years but obstructed by Nevada mining and ranching interests. When the protected land around Lehman Caves National Monument was expanded to include Wheeler Peak and redesignated a national park in 1987, the move was said to be a political tribute to a Reagan ally, retiring Nevada senator Paul Laxalt. Cattle grazing was allowed to continue in the park, a compromise that was justified by designating cows as part of the natural ecosystem.[92]

Thus even the creation of parks has political purposes and is accomplished with political compromises. The NPS admits in the Vail Agenda, "The practice of creating park units for particular constituencies has raised concerns that the criterion of national significance was being discarded and the Service's resources spread too thin."[93] Capitol Hill sources also admit that expansion has been fragmentary and politically determined, but they argue that such maneuvers are as much the result of the lack of determined leadership at the top of the agency as of congressional intervention. The most recent case involves the Presidio in San Francisco. When the army leaves at the end of 1994, the 217-year old fort will be turned over to the NPS, with a maintenance budget of more than $40 million a year and a commitment to house former Soviet president Mikhail Gorbachev's nonprofit foundation.

The best example of NPS units created for utilitarian rather than preservation purposes is Steamtown. Steamtown became an NPS site in 1986 when Scranton's own representative, Joe McDade, bypassed the stated procedure for park creation and amended the pending Omnibus Appropriation Bill. McDade exercised the power he enjoyed as the ranking minority member of the House Appropriations Committee. Because the abandoned rail yard held no original equipment and few historical structures, the proposed site came with a huge price tag. In the next year's NPS Financial Report, the extraordinary clout behind Steamtown was obvious. In the column for contractual aid was funding for eleven capital projects that were determined by statute. Steamtown USA was to receive $8 million, more than twice as much as the other ten programs combined. The creation of Steamtown has nothing to do with goals of preserving natural areas. In fact, the area may be heavily contaminated by toxic wastes. It has little to do with historic themes, either. In 1991 a panel of historians testified to Congress that Steamtown would be, when finished, "little more than a railroad theme park with an eclectic collection of trains." A former Smithsonian transportation curator called it "a third-rate collection in a place to which it had no relevance, and its determination as a historical site was simply a political trick."[94] Steamtown's creation has everything to do with economics and political clout.

By the end of the 1980s the park system was not much closer to being representative of a variety of natural regions than it had been a decade before. Research sponsored by the NPCA in 1987 concluded that even the system for classifying natural regions lacked "enough detail to be useful for surveying ecosystem diversity."[95] Even using NPS calculations, more than 40 percent of the regions were not yet potentially represented by the end of the decade. Appropriations for the study program identifying new areas were terminated in 1981, and the Office of New Area Studies was phased out during the 1980s. The NPCA has identified forty-six natural areas and forty historic sites meeting new area criteria that await inclusion in the national park system.

FUTURE.   The changes to the park creation process that occurred during the 1980s have continued into the 1990s. The Bush administration offered few initiatives on new parks. Proposals to expand existing units like the Gateway Arch and to create new units like the Presidio have been stalled by political wrangling and exorbitant cost estimates. The NPS has even begun to consider seeking corporate sponsors to assist in the financial arrangements for new parks. Gaps in the system, such as tallgrass prairies, tropical rain forests, and underwater ecosystems, have been identified, but prospects for filling them are not promising.

Because the process for creating new parks has become so responsive to political calculations, the NPS is no longer able to preserve diverse ecosystems through expansion. According to the agency's own most recent appraisal, "The general conclusion of the Vail Symposium is that the Park Service has lost some control over the process of establishing new parks."[96] If and when they are created, new park units are established because they are useful to someone, not because they represent a particular natural or historic region.

What Is Protected

Signs in Front Royal, Virginia, proclaim that the town is "where the Skyline Drive begins." Skyline Drive is the scenic 105-mile long road that runs the length of Shenandoah National Park. When park officials announced plans to close the section of road running from Front Royal to the northern entrance of Shenandoah for the winter months, the commercial interests of Front Royal were predictably upset. After all, Shenandoah draws roughly 2 million visitors a year to the area. Park officials were responding to central office demands to cut expenditures to meet a reduced 1993 budget. They claimed that closing the road during the winter months could save $200,000, which would help to avoid cuts in other programs such as poacher detection. Front Royal's commercial interests were not impressed. The executive director of the Chamber of Commerce joked that the new signs might soon read "where the Skyline Drive ends" and then warned seriously of "catastrophic" effects on local businesses. Seven members of the Virginia congressional delegation responded by sending a letter to the NPS director protesting "severe economic consequences" for their constituents. Shenandoah managers were forced to drop the road closure plan two weeks later. Representative Frank Wolf, Republican of Virginia, pronounced: "It's the right thing to do and the only solution which assures that the Shenandoah Valley's business and tourism interests are protected."[97] Wolf thereby admitted what had truly been the object of protection, and it was not the park.

PLANNING.   If you want to see the Guadalupe Mountains in Texas from the top, you hike. It is worth it. After I backpacked 2,500 feet up along the Bear Canyon Trail, I camped on top, literally the only person for miles around. The view and the quiet were remarkable. There is no choice of travel because of the planning process for parks. When Guadalupe was established in the early 1970s, its plan included a proposal for a tramway to carry tourists up into the mountains. The tramway idea was pushed by Texas representative Richard White. White refused to support wilderness designation for the park unless the tramway was constructed. Park officials inserted the tramway proposal in the master plan and then stalled until White's mind could be changed. By 1980 the park's management could report that the tramway's "development seems unlikely at this time."[98] Today, Guadalupe's chief ranger Jan Wobbenhorst calls herself "fortunate" to work in a relatively undeveloped park.

Contrast Wobbenhorst's situation with that of officials at Denali. One of the world's most spectacular and best-preserved parks is now under pressure from Senator Frank Murkowski, Republican of Alaska, to plan and build a system like the one rejected in Texas to carry tourists into the mountains. In view of the current structure of the NPS, the planning process for Denali may follow a different course than it did in the Guadalupe Mountains. Planning begins with park creation but continues regularly thereafter. Planning for each American park unit is based on two documents. The basic document is a general management plan (GMP). The GMP sets forth explicit objectives and strategies in conservation, recreation, transportation, and economic development. It is prepared by a team of planners comprising the superintendent, regional officials, and other staff members with expertise in particular areas. The GMP is to be reviewed and renewed roughly every fifteen years. Each park also uses a Statement for Management, prepared every two years by the superintendent and regional director and designed to identify major problems. Public participation through workshops, meetings, review, and comment on drafts is to be used throughout the process.

The management plans in each park revolve around zoning. Parklands are zoned "to designate where various strategies for management and use will best fulfill management objectives and achieve the purpose of the park." The NPS uses four primary zones:

—Natural: Development limited to "dispersed recreational and essential management facilities that have no adverse effect on scenic quality and natural processes" such as trails.
—Cultural: Development allowed for the preservation and interpretation of cultural resources such as historic sites.
—Development: Areas and facilities used to service park managers and visitors.
—Special Use: Subzones that include "commercial, exploration/mining, grazing, forest utilization, and reservoir."[99]

Certain aspects of the presentation of these management policies for planning are important for comparative purposes. First, the planning guidelines do not prioritize between use and preservation. They do not include a zone where access is prohibited, but they do include a zone that allows commercial uses such as mining. The language of the guidelines indicates little preference. For example, under "Evaluation of Alternatives," the section begins, "As required by the National Environmental Policy Act," and then mandates consideration of alternative approaches but recommends no action.

Second, the lack of agency autonomy for planning and zoning is explicitly acknowledged. Congress is mentioned prominently. Under the planning priorities section, the policy reads: "Congressionally directed plans will be given a priority that enables their completion within the required time frame." This mandate could easily be interpreted to include Senator Murkowski's proposal for Denali. The zoning section states that subzones will be used "as necessary to achieve the park-to-park distinctions in management emphasis called for by Congress in enabling legislation."[100]

Third, still within zoning, no attention is given to the Wilderness Act of 1964, which called for the preservation of designated pristine, roadless areas throughout the country, including those in national parks. Perhaps this inattention is not surprising given the slow responsiveness of the NPS to this legislation since its passage. Representative White's ability to withhold support for wilderness designation in the Guadalupe Mountains is a good example of why the agency is slow to respond. The pace of response has slowed so much in recent years that it has inspired a proposed bill,

H.R. 4326, "to improve the wilderness management" of the agency by creating a new wilderness chief in a planning position.[101]

The planning processes for the NPS changed little in the 1980s, but that in itself is revealing. Other national park systems modified their zoning categories and procedures to reflect an increased emphasis on preservation of natural conditions. The NPS instead retained specific use categories, legitimized congressional interventions, and was slow to adopt a renewed emphasis on the establishment of wilderness areas.

VISITOR USE.   A visit to the Redwood Information Center off Route 101 just west of Orick, California, gives a taste of the two extremes of visitor use in Redwoods National Park. I stopped at the visitor center to get my pass into the Tall Trees Access Road. To protect the forest, the NPS gives the combination code to only two shuttle buses and twenty private vehicles a day. Once into the Tall Trees area, however, you can hike into an incredible old-growth forest that is home to the world's three tallest trees. The trees, the tallest stretching nearly 370 feet into the sky, are magnificent. Returning to the center, you can also walk 100 yards down to the Pacific Ocean. Do not expect a stroll as peaceful as the one available in the forest, however. Freshwater Lagoon Beach is one of the most popular spots on the West Coast for dune buggies and other motorized vehicles. On summer days, the traffic jams on the beach are often worse than the traffic on 101. When I asked a ranger why the NPS allows the motorized sand crabs, he frowned and spread his hands: "We inherited them." Superintendent Arthur Eck was more specific: NPS jurisdiction only goes to the top of the sand dunes and not out to the water line. Thus in one place I saw an unusual restriction on visitor use through quotas and a fairly unnatural use of parklands because of jurisdictional questions.

The NPS mandate calls for the enjoyment of the parks, but in ways that do not contradict the natural experience. Enjoyment involves not just numbers of visitors but also the activities of those visitors while they are in the parks. The NPS policies on visitor use have led to both increased visitation and increasingly diverse forms of recreation by park visitors.

In its policy guidelines the NPS openly advocates increasing the number of visitors to national parks. The section on visitor use begins: "To the extent practicable, the NPS will encourage people to come to the parks and to pursue inspirational, educational, and recreational activities related to the resources found in these special environments."[102] This policy is the culmination of years of largely successful efforts to attract park visitors. At the end of the 1980s, park units were drawing more than 300 million total visits a year. National parks alone drew roughly 58 million visits. These numbers reflect a high growth rate during the 1980s. Total recorded visits to NPS sites increased by 22 percent between 1980 and 1990.[103]

Visitation is occasionally restricted through the use of quotas, as in the Tall Trees area, but the sheer volume of tourists has led mostly to procedural modifications. The most apparent change is the system for reserving campgrounds. In 1991 a toll-free telephone number was established to allow visitors to reserve campgrounds in the most popular parks. This procedure was adopted after reluctance to use any kind of reservation system led to predawn waiting lines at many popular spots. That reluctance has now been transferred to the next option, the imposition of quotas on the numbers of visitors allowed within individual parks. NPS director Ridenour cited the public reception of reservation systems as evidence that "leads us to wonder how the public would react if we tried to limit the number of visitors at some of our most popular parks."[104] Nevertheless, this is a policy that will not be adopted soon. DOI secretary Lujan openly encouraged increased visitation, calling limits "the last thing I would want to do."[105] The message to the field is clear. As Grand Canyon assistant director John Reed said when I asked him if they ever considered quotas: "Politically, we can't do that."

The NPS has allowed visitors to the parks to engage in diverse activities in recent years. Agency policy permits individual parks to allow such diverse pursuits as recreational fishing, sport hunting, trapping, off-road vehicle use, golf, and snowmobiling within park boundaries. These allowances and related restrictions are determined partly by each park's management but also by federal law. Recreational fishing is permitted in 143 units. Restrictions on catch-and-release or the use of only artificial lures vary from park to park. Sport hunting in some prominent national parks, for instance moose hunting in Grand Teton, has been publicly criticized, but the policy has not changed. Golf courses are prominent in many parks and still being planned in others. Olympic, for example, may soon get a new course built by Mitsubishi. Recreational snowmobiling has become popular at northern parks such as Voyageurs and Yellowstone.

One form of visitor use may preclude another. Since the agency is allowing increasingly diverse forms of recreation, these situations have become more common. What determines which demand prevails in a specific situation? Because the NPS is politically vulnerable, the answer is often determined by the political clout of particular user groups. Canyonlands superintendent Wickware explained to me why the park remains so inaccessible for most tourists. According to Wickware, the four-wheel drive lobby has prevented paved roads, bike paths, and hiking trails in the park so that off-road vehicles remain the predominant means of access into the undeveloped areas. The four-wheel-drivers have used their political clout, says Wickware, to prevent opportunities for other visitors, including two-wheelers and hikers. Conceivably, environmental groups can use their political clout to force specific park decisions as well. While one might argue that such behavior can facilitate preservation in specific circumstances, the impact is certainly nonsystematic and supersedes agency determination of means of attaining difficult goals.

COMMERCIAL USE.   The descendants of Senator Hansen's cows still graze in the Tetons. Today, in fact, they graze on the Triangle X ranch. Triangle X uses nearly 1,500 acres of parkland for a dude ranch. When Grand Teton Superintendent Jack Stark retired in 1991, he criticized the ranch for abusing parkland through development and overgrazing. Stark, backed by regional director Mintzmyer, proposed that the ranch's lease be allowed to expire in 1992. When asked why he had been quiet before, Stark admitted that some things are better left unsaid until retirement. A subsequent DOI investigation confirmed some of Stark's allegations, such as dumping and oil contamination, but concluded that overgrazing could not be confirmed until after a three-year study, which would thus end after the lease had been renewed, and that there was "no lasting damage to the resources of the park."[106] These findings should not have been a surprise. The Triangle X is owned by the Turner brothers, one of whom was serving at the time of the decision as director of the Fish and Wildlife Service.

Parks have been used by commercial interests, in addition to concessioners, within and near park borders. As the discussion on zoning indicates, commercial use of park resources through special uses is still officially allowed. Special uses include such activities as logging, mineral extraction, and grazing. The official policy on these activities allows them inside park boundaries if operators obtain legal permission from park officials. One might argue that this NPS policy would allow a preservationist simply to refuse to give the required written permission for special uses. However, the wording of the policy is revealing. First, special uses require "written permission from an NPS official in order to take place." Since some NPS officials are political appointees, this wording means potential users can bypass the employees they know are preservation-minded. Second, the policy carefully notes that "use may be permitted only if the activity has been judged by the superintendent not to cause any derogation of the values and purposes for which the park was established." Since American parks are created by individual pieces of legislation, each act describes the purposes for which the park was established. This tradition dates to the second amendment of the 1916 National Parks Act, which "authorizes the Secretary of the Interior to grant the privilege of grazing livestock in any national park, monument, or reservation, except Yellowstone."[107] When NPS superintendents have tried to apply the policy strictly, they have often found it necessary to rely on higher levels of government to support them. In that sense, the policy has not changed since Shanks challenged Senator Hansen's cows.

One recent policy change encourages a different kind of commercial use of the parks. In 1991 the agency reversed a previous ban on permanent commercial displays in parks to allow recognition of gifts from corporate donors. While the agency plans to confine such displays to small items such as plaques, the policy change opens the door to extensive commercial activity.

Another recent policy change was proposed but not enacted. This proposal would have prohibited geothermal drilling on the borders of national parks, particularly Yellowstone. The proposal, sponsored by Representative Pat Williams, Democrat of Montana, was inspired by the attempts of the Church Universal to drill a well within two miles of Yellowstone's borders, an activity that might easily affect geothermal activity inside the park. Congressional action on the proposal was, in part, precluded by the Department of the Interior's recommendation to allow the well. In the DOI recommendation, Secretary Lujan included a U.S. Geological Survey report citing "no discernible risk" but did not include strong warnings from Yellowstone superintendent Robert Barbee nor his agency's twelve-page criticism of the project.[108]

The NPS has an official policy toward at least one other commercial use of park space, airline travel. Though the agency admits that park space is affected by aircraft, particularly through noise pollution, its own policy calls for monitoring of effects but, "since the NPS has no direct authority or jurisdiction over airspace above parks," nothing stronger.[109]

FIRE MANAGEMENT.   In the summer of 1988, with much of his park in flames and thousands of firefighters battling the blazes, Yellowstone superintendent Robert Barbee attempted valiantly to defend the existing let-burn fire policy of the NPS. Barbee, appearing weary and worn by the events of the summer, said "This may not be convincing to anybody, but here it is: The role of a national park is to be a repository for natural processes."[110] Within weeks, that NPS policy and more than one-third of the park was in ashes.

The NPS has policies for the management of fire. How often are park managers to suppress naturally occurring fires? Should managers use prescribed, man-made fires to facilitate ecological change? Fire policies were in transition when the formal guidelines in *Management Policies* were promulgated in 1988. As fires raged in Yellowstone that summer, the agency published the following note: "Fire management policies are under review . . . and will be modified as necessary."[111] To understand NPS policy on fire, then, requires understanding the Yellowstone situation. The fire policies used at Yellowstone have traditionally been copied throughout the system. What rose from the ashes of the 1988 conflagration would thus be the fire policy for the entire agency.

Fire policies at Yellowstone and throughout the NPS had changed once already before 1988. Fire policy before 1972 had been "one of total suppression based upon interpretation of the enabling legislation to 'preserve'." Between 1972 and 1976, fire policy shifted to allow "fire to play its natural role in the park and thus perpetuate natural ecosystems."[112] The natural burn or let-burn policy enabled managers to allow naturally occurring fires to burn as long as they did not threaten human life and property. The change came partly in response to the 1963 Leopold Report, which encouraged the return of parklands to natural conditions, and partly in response to new views of environmental management. As the 1973 Master Plan for Yellowstone states: "Yellowstone should be a place where all the resources in a wild land environment are subject to minimal management."[113] When the park did fight fires, it would try to minimize the use of heavy equipment such as bulldozers. The changes did not go as far as some would have liked. Calls for the use of prescribed fires, based largely on research on past Indian practices, went unheeded. Yellowstone refused to adopt such procedures in part because of "mistakes" in applying the technique in other areas.

The summer of 1988 was the driest in Yellowstone's century of record keeping. Dry fuel in the forests had built to dangerous amounts. Of 235 let-burn fires between 1972 and 1987, 205 burned less than an acre.[114] Park managers continued to resist using prescribed fires. As table 2-9 shows, few of the 2.2 million acres of the park had been burned in decades. Summer arrived with more than twice the normal amount of lightning and with winds gusting up to seventy miles an hours. By mid-July, roughly one percent of the park's acreage was on fire. Park naturalists and officials tried to maintain and defend their let-burn policies despite increasing calls for control from local concessioners and developers and their congressional representatives. On July 21, after a visit by DOI secretary Hodel, policy changed to mandate suppression of all fires. Nevertheless, the usual August rains never arrived, and the fires grew. By late summer, with millions watching on television, more than 25,000 firefighters were battling forty different fires. The fires came so close to Grant Village and Old Faithful that firefighters were hosing down the roofs of buildings, when an early September snowfall extinguished the blazes. As the figures in table 2-9 show, roughly 36 percent of the park had burned.

The fallout from the fires was intense. As one observer wrote, "The 1988 Yellowstone fire stimulated greater interest in national parks management than any event in recent times." Many park managers and scientists continued to defend their policies and claimed that, in many ways, the fires had been "an ecological boon."[115] Lands were rejuvenated, natural processes were restored, less than 1 percent of large mammals were killed by fire, and ground was not destroyed since the fires burned through

TABLE 2-9. *Acres Burned in Yellowstone National Park Fires, 1930–91*

Year	Fires	Acres	Year	Fires	Acres
1930	22	101	1961	37	658
1931	81	20,519	1962	30	4
1932	30	2,050	1963	46	38
1933	40	2,446	1964	20	16
1934	39	667	1965	12	54
1935	56	103	1966	44	292
1936	57	26	1967	20	3
1937	41	4	1968	16	2
1938	12	1	1969	18	6
1939	35	1,775	1970	14	2
1940	64	20,115	1971	24	21
1941	12	2	1972	21	3
1942	48	1,022	1973	32	151
1943	27	550	1974	38	1,308
1944	5	63	1975	26	6
1945	4	1	1976	30	1,603
1946	29	838	1977	29	68
1947	13	4	1978	24	16
1948	27	84	1979	54	11,240
1949	56	2,988	1980	25	4
1950	37	0	1981	64	20,575
1951	8	1	1982	20	1
1952	28	72	1983	7	0
1953	81	1,760	1984	11	0
1954	18	54	1985	53	32
1955	17	33	1986	32	1
1956	19	200	1987	35	964
1957	21	27	1988	45	794,002
1958	29	80	1989	24	11
1959	20	6	1990	43	247
1960	55	683	1991	29	265

Source: U.S. National Park Service (1992d, appendix C).

rather than down the trees, leaving many trunks still standing. On the other side of the issue, local economic interests and politicians were harshly critical. Superintendent Barbee was ridiculed as "Bob Barbecue." Wyoming senators Simpson and Wallop denounced NPS policies as a "disaster" and demanded the resignation of NPS director Mott.[116] As early as the fall of 1988, even park managers recognized the inevitability of changes in their fire policies. Yellowstone research manager John Varley admitted, "Of course we're going to revise our policy. We just took a giant step forward in understanding fire behavior. That policy has evolved."[117]

The fact that those changes were to be politically determined soon became apparent. The following exchange took place in the 1988 Senate hearings.

> Senator Baucus: "The important point is not to call for the heads of people, to be partisan, to be political, or rhetorical, but rather to get to the bottom of the matter here and see if there is a way to help us constructively reshape the prescribed burn policy in a way that makes more sense."
>
> Senator Wallop: "I suppose they have not been political. I wish they had not been. The groups that have been attacking me, National Parks and Conservation Association, the Sierra Club, . . . are all east and out of the picture of the immediate environments of your states and mine. So if we do not want to make it political, leave us not."[118]

The dialogue illustrates just how political the decisions had become. Policy alterations involved interest groups, politicians, and especially concern for immediate constituents. As one Yellowstone official told me succinctly, "Washington made us rewrite our fire plan."

The agency's procedure to change policy began in December of 1988 with an interdepartmental review by officials from the NPS, the Bureau of Land Management, the Bureau of Indian Affairs, the Fish and Wildlife Service, and the Forest Service. Other public land agencies were included because the fires had affected numerous areas, particularly national forests, around Yellowstone. This was an important development since the other agencies, led by the Forest Service, had traditionally suppressed natural fires. The Interagency Agreement for this review team makes the tone apparent in the first sentence: "Fire loss in the forests and on the rangelands of the Nation continues to be a matter of great concern to the American public."[119] In listing objectives and goals, the use of fire to restore natural processes is conspicuous only for its absence. Most critics of NPS policies forgot or ignored the contribution to the fires of the accumulations of dry wood that had resulted from having too few planned fires.

The review team's first recommendation was to adopt an interim policy of fighting all natural fires until the review was completed. Still, the report did not recommend a complete abandonment of let-burn practices; it concluded that the "objectives of prescribed natural fire programs . . . are sound, but that policies need to be refined."[120] This assessment provoked a new round of criticisms from developers and politicians. Idaho senator Steve Symms said, "I doubt the public will be as gentle with the let-burn policy as the review team was."[121]

The fire policy that resulted from review team recommendations and other considerations is somewhat complex. The moratorium on let-burn natural fires continued through 1989 and into the summer of 1990. Yellowstone's *Fire Management Plan* divides the park into zones: suppression, where all fires are fought; conditional, where fires are allowed to burn if they meet certain conditions; and prescribed natural fire, where natural and prescribed fires are allowed to burn if they are not harmful. The park's 1991 management plan states somewhat ambiguously: "Allow fire to play its natural role in the ecosystem while protecting human life, facilities, and cultural resources."[122] In both documents, the emphasis on protection of human life and property is obvious. Parks throughout the NPS system adopted Yellowstone procedures: land managers must issue daily reports on fires and must fight them if they cannot guarantee that the blazes will remain confined to appropriate areas.

The result of this debate is that the NPS has an implicit fire policy that is much more restrictive than the policy that existed between 1972 and 1988. Under the new policy, every day that a superintendent allows a fire to burn, he or she must sign documents stating that the fire meets certain requirements. The message is clear. Managers can let natural fires burn and can use prescribed fires if necessary, but they do so at their own discretion and at their own risk. Stephen Pyne, a Grand Canyon firefighter for two decades, described how employees are caught between the protection of people and property and the use of natural fire. As Pyne noted, protection usually wins out: "Every Park Service program is designed to please the visitor and encourage good feelings about the Park Service. The actual natural resources of the park are only a means to this end."[123] Pyne's opinion was echoed in assessments of the new fire policy by NPS management and by political authorities. Natural resource director Hester told me, "When Congress holds a hearing on fire, they want to know what the effect is on the visitor." Senator Wallop assessed the final policy that resulted from the Yellowstone discussions: "All the words about natural fire are in there, but the fact is they're now going to have to suppress the fires. You and I know that no bureaucrat is going to sign a certification that he will not be outwitted by God and a wildfire. Instead, they'll have to fight those fires."[124] Sandy Rives, a Shenandoah official, reinforced Wallop's appraisal of the ultimate effect of the Yellowstone fires: "Superintendents now say, 'By God, I just can't take a chance.'"

SCIENCE AND RESEARCH.   The purest lake in the United States and possibly the world may lose some of its clarity as a result of inadequate scientific data. Crater Lake was formed when a mountain collapsed in on itself. All of the water in the crater thus results from snow and rain. No streams or rivers carry impurities into Crater Lake. The water contains about one-sixth of the particles that are allowable in a unit of tap water. The lake is the seventh deepest in the world (1,932 feet). As a result, the clarity of the water is unsurpassed. In the depths of the lake, scientists have been able to see a secchi board (which looks like a small black and white TV test pattern) at 140 feet. These factors explain the remarkable blue color of the water. As light passes through clear water, other colors are absorbed, leaving only the blue. The clarity and the purity have all been threatened recently. In 1989 a company proposed drilling a geothermal well just outside park boundaries. NPS officials feared that the well could strike a vent or shaft connected to the lake bottom, thereby warming the water, creating bacterial mats, producing algae, and reducing clarity. A court ruled that the NPS had to prove that Crater Lake already had geothermal activity that might be connected to the proposed well by finding areas where water was ten degrees or more warmer than the rest of the lake. The lack of scientific data on geothermal activity in the area almost killed attempts to stop the drilling. Since geothermal vents can be as small as the size of a manhole cover, the agency ultimately rented the Deep Rover used in the movie *The Abyss* to search the bottom of the lake. As one ranger told me, "We were all on pins and needles waiting for the daily report from Deep Rover." One day, the divers found a seventeen-degree difference, and the well project was temporarily shelved. The hold is only temporary because, as one official said, "There's big money in this," and because the park has yet to establish systematic proof that increased geothermal activity would harm the lake.

Historically the NPS has given little more than lip service to the idea of consistent policies on science and research. One reason is that research efforts were not systematic but instead were dependent on the personalities of key figures involved. When those key people were lost—the tragic death of George Wright in an automobile accident, for example—commitment was lost with them. The lack of continuity in research emphases prevented the long-term commitment needed for scientific efforts. A second reason involved the traditional decentralization of the organization. Research decisions, but not necessarily supporting funds, were left to individual park superintendents. Without funds, without the resources for coordination across parks, and often without motivation, research efforts died at the field level. A third reason is that scientific reports are subject to political manipulation. In the recent Yellowstone geothermal drilling case, for example, NPS concerns about the scientific uncertainty of the Geological Survey study were not included in the final report to Congress.[125]

In recent years, the NPS has shown greater concern about this issue but not necessarily greater commitment to changing policies on science and research. Commitment to scientific research became official policy in 1988. In its policy guidelines the NPS states: "A program of natural and social science research will be conducted to support NPS staff in carrying out the mission of the National Park Service by providing an accurate scientific basis for planning, development, and management decisions."[126] The NPS announced a five-year program to focus funds where needed, to improve research training of personnel, to improve park information gathering, and to improve dissemination of research results. The agency further identified research needs in a variety of important issues. For example, the effect of exotic plants on indigenous species was identified but not well understood at eighty-eight parks. Another 1988 project identified over 120 endangered or threatened species that could be found in more than 140 units of the system. The agency directs its science efforts at these in-park issues: "The science program will be focused on applied research necessary to direct management actions in pursuit of park objectives as stated in legislation and planning documents."[127]

These policy statements are somewhat vague with little explicit discussion of funding levels or systematic procedures. The focus on in-park projects tempts emphasis on short-term, specific, applied issues that can be politically determined. Several recent assessments suggest that commitment is still lacking. An NPCA advisory commission concluded in 1989 that the agency's research program consisted of mainly temporary, underfunded attempts at putting out brush fires rather than efforts at pure research. Another critical report was issued by the National Academy of Sciences in late 1992. The report concluded that the NPS suffers from a lack of basic science and long-term research efforts. The study points out that only 2 percent of the NPS budget now goes to research. Finally, the agency's own assessment of its commitment to science and research is harsh: "The NPS is extraordinarily deficient in its capacities to generate, acquire, synthesize, act upon and articulate to the public sound research and scientific information." The agency report reiterates the need for political support and clear goals by admitting that efforts that have begun are often uncoordinated and narrowly focused because "the Park Service lacks specific legislative mandates supporting a system-wide commitment to science and an integrated program of resource management and research."[128]

EDUCATION.   A visitor to Manassas Battlefield, less than an hour outside Washington, D.C., can receive an inspiring education in how the American nation struggled with its own conscience and also a disturbing education in how the NPS is forced to preserve that heritage. The site of two important battles in the Civil War, Manassas has an informative visitor center and an educational hiking trail through sites where more than 20,000 men fought and died in the early 1860s. Manassas also hosts a stable that had nothing to do with the battles but is home to nine horses that provided recreational riding for Vice-President Quayle's family and other dignitaries. In a late addition to a 1993 NPS budget that was already $34 million short of forecast needs, Manassas received $38,000 for expansion and improvement of the stables. Park managers did not request the money. The Manassas superintendent did request money to improve educational efforts, including $13,500 to provide disabled access to a map room in the visitor center and $30,000 for preventive maintenance on historical buildings. While the park did not receive a dime of those requested funds, the park employees will have nice stables. Unfortunately, only three of the park rangers even know how to ride. Overriding the past objections of members of Congress, Interior secretary Lujan directed the NPS, in a December 4, 1992, memorandum, to complete work on the stable as soon as possible.[129]

The NPS has the opportunity to educate the public about historical and ecological issues. The need to inform and educate visitors has become even greater in recent years as the parks face increasing threats to their historical integrity and to their ecological health. NPS policy calls for using "interpretive programs in all parks to instill an understanding and appreciation of the value of parks and their resources; to develop public support for preserving park resources; to provide the information necessary to ensure the successful adaptation of visitors to park environments; and to encourage and facilitate appropriate, safe, minimum-impact use of park resources."[130] The focus in that statement and in the accompanying guidelines is on using interpretive programs to educate visitors about the resources and activities within the park units. The NPS uses outreach services to disseminate information to those who cannot physically visit the park, but, like applied scientific research, the information is focused on the park itself.

The American battlefields have been particularly beleaguered in recent years. In 1986 developers purchased a tract of land adjacent to Manassas on which to build a development of homes, shops, and offices. In this regard, the proximity of Manassas to the nation's capital was a blessing. Preservationists and Civil War historians protested enough to inspire Congress to repurchase the land to provide a scenic buffer zone for the park. Antietam, also close to Washington, has received similar special congressional protection. Other battlefields, more removed in location and attention, have not been so fortunate. Even the most prominent historical sites have not been spared the pressures of economic development. Today, looking across the field of Pickett's Charge at Gettysburg, one's sense of awe is disturbed by the sight of a 300-foot observation tower. The tower was built despite the objections of historians, NPS employees, and the government of Pennsylvania. When I talked to former Gettysburg ranger Jerry Schober in 1992, he was still bitter about the tower. As Schober says, the tower was built for economic reasons, not for educational purposes.

In the area of environmental education, agency personnel are their own toughest critics. The Vail Conference recognized the role the NPS can play in educating the public not just about the park itself but also about the lessons that can be learned from natural preservation. The conference decried the inconsistent commitment "to a mission of proactive education and interpretation," attributing this inconsistency to the continued existence of a "fortress mentality" wherein employees "too often focus their attention only on what happens within the confines of national parks. Issues are defined in terms of what they mean for a particular park site, rather than the larger ecosystem, nearby communities or a region as a whole."[131] The NPCA Advisory Commission on Research and Resource Management also concluded that the agency needed a much greater emphasis on educating the American public about "the importance of our parks."[132]

This internal focus is an inevitable result of the lack of institutional respect given the agency. Park employees have a hard time trying to educate the public about ecosystem management when the politically motivated bureaucrats managing surrounding lands maintain hostile relations with the NPS. The resulting inability to increase public awareness of ecosystem issues makes it difficult, as the Vail conferees concluded, for the parks to meet the external threats they recognized as early as 1980.[133] As Yosemite superintendent Finley said to me: "Isn't it interesting that nobody's updated that document [State of the Parks]? Have these threats gone away in the last twelve years?" Without proactive education, these threats will not be addressed.

Parks in West Virginia Always Get Enough Money

In two separate interviews more than a year apart, the comptroller general of the NPS repeated the same answer when I asked him about political influence on the budget process: "Parks in West Virginia always get enough money." The point is obvious: the chairman of the U.S. Senate Appropriations Committee, Robert Byrd, is from West Virginia. Once the parks are created and the policies formulated, the agencies need resources to manage the sites. Politicians such as Byrd are determining priorities in the use of resources, human and financial, as well as in the selection and management of parks.

PERSONNEL.   The NPS history of the past two decades is full of stories of preservation-oriented personnel who chafed at how they and the parks were used. Al Runte, author of several books on the parks, was reprimanded for giving interpretive talks critical of concessioners in Yosemite. Art Sedlak of Glacier was suspended and then immortalized as a folk hero for shooting an illegal snowmobile. Bernard Shanks worked in six national parks but never forgot his interaction with Senator Hansen's cows. Veteran rangers Chapman and Mintzmyer were driven into retirement for advocating preservation-oriented policies. While the individual circumstances of each case are unique, all exemplify employee frustration with an overall orientation toward use of the parks at the expense of preservation.

The NPS employs roughly 12,000 full-time employees. Add to that figure the seasonal workers (four three-month seasonals equals one full-timer), and the full-time equivalents total about 16,000. The pay, training, and career paths of these employees suggest how little attention is given to preservation goals with personnel and why so many cases like those cited above have occurred recently.

The low compensation of these employees contributes to several personnel issues that make emphasis of preservation difficult. First, hiring qualified people has become harder in recent years. More than a third of new hires lack college degrees. Of those with degrees, only half hold them in subjects related to park management. Only 40 percent of agency professionals hold degrees in resource management fields. The agency eliminated formal educational requirements in 1969, and "ranger educational attainment has been steadily declining since."[134] Second, retaining good people has become problematic. As mentioned before, turnover rates among NPS employees have more than doubled in the past two decades. Agency employees are increasingly lured to private jobs or even other public land agencies that pay considerably more. Other public land agencies have better reputations for providing career opportunities. As Big Bend's Arnberger told me, "The Forest Service does a much better job than the NPS in promotions." Third, those employees who stay are often stuck in low-paying positions because the work force has constricted in recent years. As a result, morale suffers. Together, these factors prevent the long-term commitment and dedication that are necessary to achieve preservation goals.

Training of NPS careerists has little to do with preservation. The NPS has three training centers: one at Albright in Grand Canyon, one in Harper's Ferry, and one in Georgia, which is used for law enforcement. The agency uses no systematic pattern for sending personnel to these training facilities, but application for courses is often up to the employees. An instructor at Albright told me interest was high; on average, four candidates apply for every slot in a class. Courses concentrate on ranger skills and law enforcement; little attention is given to scientific research or resource management. Most employees thus start out without professional degrees and then receive little in the way of resource training. "The consequence," as the Vail group points out, "is a work force with declining education and professional qualifications at the very time when far more is needed."[135]

The typical career path for the NPS employee illustrates both the emphasis on visitor service and the centralizing tendencies in the agency. The projected path involves eight to ten years in the field, the next ten years assuming supervisory positions, and then years after twenty in managerial positions as superintendents or regional officers. While the first ten years could conceivably be spent in resource management or other preservation-oriented positions, in fact a vast majority of managers come from visitor protection. A mid-1980s poll by the Association of Park Rangers showed eleven times as many managers came from visitor protection positions as from resource management. This does not necessarily indicate that these managers will ignore preservation, but often "resource management is neglected in favor of visitor protection duties, such as law enforcement."[136]

The NPS admits that many managers are thrust into such roles with little formal preparation. Centralization pressures increase as a result. The Vail document states: "The fact that many superintendents are poorly prepared for their managerial roles invites regional and headquarters managers to closely monitor their performance, impose bureaucratic controls and rob young superintendents of their autonomy and opportunity to grow."[137]

Despite recommendations to the contrary, recent cases reveal a decreasing emphasis on preservation. The cases illustrate how field preferences are overruled. First, in the early 1980s a task force of rangers and resource managers was asked to consider standards for the ranger corps. The task force responded with recommendations for academic requirements that are more focused on natural science and resource management. Both Director Dickenson and Director Mott overruled the recommendations, the former saying the recommendations were too restrictive. According to one NPS official, "He wanted more freedom to make appointments than the task force recommendations would allow."[138]

Second, the Grade Comparability Task Force created in response to the 1986 *Twelve-Point Plan* considered career paths. The task force recommended replacing the current career path with a multiple-track arrangement whereby more field employees from the resource management areas could compete with those in visitor protection for supervisory promotions.[139] The recommendations have yet to be made effective.

Third, consider the effect of the Chapman and Mintzmyer cases on the perceptions of field employees: two career veterans lose their jobs for opposing the forces advocating use as the prime mission of the agency. The cases were highly publicized, and NPS employees noticed. As one

TABLE 2-10. *Budget Allocations of the NPS, 1990*
Millions of dollars unless otherwise specified

Item	Percent	Allocation	Appropriation
Maintenance	23	277	1,193
Visitor service	16	188	1,193
Resource management	10	125	1,193
Park supervision	6	73	1,193
Research	3	30	1,028

Source: U.S. Department of the Interior (1990a, NPS-12, 29, 94–102).

young superintendent told me, "It makes me sick." It may also make him change subsequent decisions.

Finally, in 1991 DOI officials recommended giving departmental political appointees power over the final steps in NPS career paths—power to approve all transfers and appointments at the GS-fourteen level and above. Regional managers and nearly all superintendents would have been affected. This most recent attempt renews the efforts that were made throughout the 1980s to increase political power over high-ranking careerists. Former director Hartzog called it "a continuation of the politicization of the Park Service."[140] Given the emphasis of political authorities on use rather than preservation of the parks, career-minded personnel realize that such a procedural move would inevitably alter their own behavior.

BUDGETS.   One of the standard jokes in the NPS is to refer to the agency as "the National Pork Service." When I heard this remark, I thought of Comptroller General Sheaffer's comment about West Virginia parks. When I returned to Washington, I asked him about congressional pork. He responded, "Much of our money is going to the wrong places for the wrong reasons." The NPS does not receive its revenues directly but rather as part of the appropriations processes. As a result, political authorities have considerable input into how these resources are to be used. Their recent spending decisions reveal priorities other than preservation.

Table 2-10 displays the allocation decisions of the NPS by function. The table shows the latest available figures, mostly from fiscal 1990. The numbers in the table reflect percentages of total expenditures by the NPS. The numbers for allocation are the actual figures in millions. For example, the NPS spent about 23 percent (277 million out of 1,193 million) of its 1990 budget on maintenance. I also calculated the percentages in adjacent years to ensure that these figures were representative. The percentages were virtually identical. The percentages do not total 100 because of all the other categories such as headquarters staff and public relations that are not included.

The NPS research figure obviously uses a different base and thus warrants some discussion. The NPS does not list research as a separate item in its budgets. Thus the fiscal year 1991 budget does not show historical records for this expenditure item. Therefore, I went through the budget justifications and added up the research items to get a total of $29,658,000 for proposed research projects. This may be an overstatement, since the NPS has no guarantee of receiving this much in actual funds. That total was then divided into the total 1991 appropriation to get a 2.9 percent figure. I am confident with this figure. Other analyses have consistently listed the research total as less than 3 percent, one study showing only 2.3 percent in fiscal year 1987.[141]

The table reflects the prioritization of visitor service. If anything, these figures are understated. Much of the money in other categories not included in the table (such as public relations and communications) could easily be counted as expenditures for visitor service. Gene Hester, NPS director of natural resources, put the actual percentage spent on things related to visitors closer to 80 percent. Nevertheless, the pure visitor service category and maintenance together account for four times as much spending as resource protection. Further, the 3 percent allocated to research is far less than the 10 percent recommended by the National Academy of Sciences.[142]

Table 2-11 analyzes allocations to individual park units. The analysis uses an econometric method called Ordinary Least Squares (OLS). For those unfamiliar with OLS, the method involves multiple regression analysis of the effect of different independent variables on one dependent variable. OLS holds other independent variables constant while examining the relation between one potentially influential factor and a dependent variable. Two summary statistics are obvious. First, the R^2 number at the bottom of the columns represents a rough approximation of how much of the variance has been explained by the independent variable. In this case, the numbers are high enough to indicate that the equation provides substantial explanatory capability. The second summary statistic requires more explanation. The numbers in parentheses are t-statistics representing the coefficient of each variable divided by its standard error. Again, for those unfamiliar with OLS, a rough guide says that if a t-statistic is greater than two, then the relationship between that independent variable and the dependent variable is significant with less than a 5 percent chance of random occurrence. All variables are significant.

The dependent variable for the equation is the amount of federal money appropriated to each individual park unit. The independent varia-

TABLE 2-11. *OLS Estimation of Park Unit Appropriations*

Independent variable	Coefficients
Constant	−4.53[a]
	(2.16)
Visitors	0.54[a]
	(13.25)
Years	11.41[a]
	(3.34)
Size	2.14[a]
	(2.65)
Threat	5.84[a]
	(8.03)
N (number of cases)	279
R²	.52

Sources: Dependent variable (FY90 expenditures), U.S. Congress, House Committee on Appropriations (various pages); visitors (visits per year), U.S. Department of the Interior (1990a, p. NPS-12); years (years before 1990 park estimate), U.S. National Park Service (1989b); size (total gross acreage), U.S. Department of the Interior (1990a, pp. NPS 222–39); threats (to park units), U.S. National Park Service (1980b, pp. 9–10).

a. Significant at the .05 level.

bles represent the number of visitors to each unit, the years since the unit was established, the size of each unit, and the number of threats (such as erosion or pollution) facing each unit. If number of visitors alone determined allocations, then that variable alone might dominate the equation. Or threats might dominate if preservation were the sole emphasis in allocation decisions. The equation is quite strong with multiple significant independent variables. Two variables that are significantly related to the allocation of funds to park units are visitors and threats. These relationships suggest that unit allocations are determined by how many visitors a park entertains in a year *and* the number of threats it must manage.

Two factors may explain the apparent lack of evidence for the hypothesized shift toward use. A statistical reason for inconclusive quantitative evidence in table 2-11 has to do with multicollinearity in the equation. Multicollinearity is a term that refers to the effects of presumed independent variables on each other. Threats are often related to number of visitors. In many cases, when visitation is high, so too are the number of threats. Separating the effects of the two is thus difficult. To try to address this, I tested each equation while omitting the variables for visitors or threats with results showing similar significant relationships. This method still does not correct the fact that large numbers of visitors create threats. Of course, the NPS must address visitor service. How much is spent on preserving the parks from other threats is difficult to determine in this analysis. A substantive reason for inconclusive quantitative evidence for the hypothesis is that individual cases of obvious political influence are buried within statistical analyses. Visitor service is a narrow measure for the use side of the mission statement. One need recall that American parks are used for explicitly political purposes in addition to just visitor service.

The Steamtown example illustrates political determination of expenditures. Between 1986 and 1991, more than $40 million was appropriated for design and construction of the historic site. Now half finished, Steamtown is like a long-term defense contract. If Congress stops funding it, the funds spent thus far will have been almost wasted. The superintendent warns that if Congress cuts even $20 million out of future appropriations, the two planned museums may not even have bathrooms. If the project continues as planned, the site will include the museums, a new visitor center, a restored railroad yard, numerous antique trains, an operational train turntable, a theater, and the potential for taking excursion rides into the Poconos. The final price tag will be at least $63 million, according to the General Accounting Office. That total does not even include such figures as $12 million for restoring railroad cars. To put this in perspective, the annual operating appropriation to Yosemite is less than $15 million.[143] Recent calls for congressional cuts in Steamtown appropriations have galvanized the local community in opposition. When I visited the site, I heard no apologies whatsoever. The site's own newspaper begins its 1992 summer issue with: "If you thought it was interesting to wind your way through construction activities over the past two summers, 'you ain't seen nothing yet' . . . [with] almost a dozen construction projects underway simultaneously."[144] When an interpretive ranger admitted to me he had heard complaints of pork barrel funds, a visitor nearby chimed in, "Yeah, but we need the money in this area." They continue to get it. The 1992 NPS budget included $13 million for Steamtown.

Steamtown is hardly the only recipient of NPS pork. The application of NPS funds to congressionally approved pet projects has historical roots but has become rampant recently. Big Bend superintendent Arnberger assessed recent changes: "The federal budget is becoming more of a line-item process wherein members of Congress take care of pet projects." Pork barreling was so obvious in 1990 that several agency career officials complained to their political bosses in a memo that the NPS was becoming a "repository for what are in essence economic development type projects."[145] The projects included:

—"America's Industrial Heritage," sponsored by Representative John Murtha, Democrat of Pennsylvania, to the tune of $13 million for industrial sites (such as the first aluminum factory) in nine Pennsylvania counties;

—a new visitor center at Fort Larned, Kansas, pushed by Senator Robert Dole, Republican of Kansas, at a cost of nearly $600,000 even though the NPS had said the existing facility was adequate;

—a $4.5 million renovation of the historic Keith-Albee Theater in Huntington, West Virginia, at the behest of Senator Robert Byrd, Democrat of West Virginia, even though the theater had been rebuilt into a four-screen multiplex cinema.

Indeed, West Virginia had received its share of the pork barrel, as Comptroller General Sheaffer knew it would. These were just a few of the projects added to the NPS appropriation at the last minute and without public hearings or scrutiny. This loophole in expenditure allocations continues to be popular. The 1992 agency budget contains $33 million in projects not authorized through the regular appropriations channels.[146]

The allocation of funds for political pork is especially problematic because the money is being sliced from a shrinking budget pie. Congress cut the 1993 NPS "minimum needed" budget request by $34 million to $984 million, just $39 million over the 1992 figure in nominal terms. With demands and visitation still increasing, the NPS has been forced to cut projects, to reduce service, and even to close certain areas. For example, Shenandoah and Yosemite will both close campgrounds in 1993, and Olympic will staff only one of four entrance stations. The agency has been forced to rely increasingly on volunteers to pick up the slack (and the trash) in numerous areas. Some officials are considering the possibility of charging people who are rescued from hazardous situations for the costly search and rescue operations.[147] In 1992 alone, the NPS spent nearly $5 million on several thousand rescues throughout the system. If people who were rescued were charged, they would just be the latest users of parks to be affected by the political manipulation of agency resources.

Summary

Changes in the political environment of the NPS over the past fifteen years have altered the agency's operating environment and determined its policy behavior. This change in NPS priorities is widely recognized and often lamented. Many in the NPS recognize their agency's new tendencies. To their dismay, they seem powerless to change them as they trudge along in the tracks of the bulldozer.

Chapter 3

Choosing Destiny

First superintendent: "Used to be we'd come here and each say our bit about our park and then the DG [Director General] would tell us what to do anyway."

Second superintendent: "Heck, used to be we'd come here and the DG would just tell us what to do."

—*Conversation overheard at Prairie Region staff meeting, 1992*

The point of the story, as Prince Albert superintendent Dave Dalman told me, was that now superintendents would come and say their piece, and the director general was actually interested in points of view from the field. As the story suggests, Canadian field managers are moving in front of the symbolic bulldozer. When trail-level bureaucrats have real power, they are often able, in attempting to preserve natural areas, to stop the bulldozer.

Unlike its American counterpart, the Canadian Park Service (CPS) since 1979 has enjoyed a growing consensus around its preservation goal and a more supportive political structure. Employees have responded by embracing their institutional autonomy, decentralizing responsibilities, and empowering trail-level bureaucrats. CPS decisions and actions are shifting to reflect the preservation focus of the agency's mandate.

The Agency before Change

Much of the early history of Canadian national park management was comparable to that of the U.S. National Park Service (NPS). Though Canada preceded the United States, indeed the world, in the creation of a national parks agency, often Canadian policymakers consciously attempted to copy the American experience. When Canada established Banff as a reserve in 1885, for example, it used American parks as models. Yellowstone had been created thirteen years earlier as "a pleasure ground for the benefit and enjoyment of the people"; Banff was to be "a public park and pleasure ground for the benefit, advantage, and enjoyment of the people."[1]

To consider how to use a park to attract tourists, Canadian officials visited and studied Arkansas's Hot Springs in 1886. Several other reserves were set aside before an oversight agency was established, the same pattern that had developed in the United States. When the Dominion Parks Branch was created in 1911, it was placed within the Department of the Interior, a move that foreshadowed later American developments. The legislation establishing the parks agency, the Dominion Forest and Reserves Park Act, was largely a product of the growing North American conservation movement, which also affected the NPS. The two park systems had similar goals and experienced similar pressures. Canadian officials and observers continued to analyze American management practices in their own agency's formative years.

However comparable to that of the United States, the Canadian parks agency earned its own particular reputation during the twentieth century. That reputation was fostered by strong economic pressures and political receptivity to those pressures. The Canadian park system was shaken by several developments of the 1970s, but overall the system retained a reputation into the 1980s for managing parks to be useful more than to be preserved.

The Focus of Economic Development

Most Americans are not quite prepared for what awaits them in Banff. My father and I had been camping in the rainy mountains for several days before we drove down into the town that takes its name from the nearby hot springs. We found much more than hot springs and warm showers. Banff is a city of more than 7,000 residents, sprawling beneath and into the snow-capped peaks of the park. Add to this nearly 4 million visitors a year, and the result is intense congestion, pollution, and traffic jams. Even experienced CPS employees shake their heads when asked about the town. Perry Jacobson, an eighteen-year Banff resident and now chief warden at Kootenay, describes the history of the town as "just like a cancer—it just keeps growing." Many observers agree that excessive development "reaches its zenith in Banff."[2]

Banff is the most striking legacy of an early emphasis of Canadian parks management. Canada's national parks were initially designed and developed largely for economic purposes. The parks were established to provide for tourism and recreation, often while allowing commercial development and resource extraction. One analyst summarized this appraisal explicitly: "Most Canadian national parks have not been removed from economic development, but have been the focus of that development."[3] Government policies toward the parks emphasized these goals for decades and still occasionally influence their management as "historical policy relics."[4]

TOURISM.   The goal of attracting tourists dominated parks policies for decades. Tourism has long been an important industry in Canada, historically constituting roughly 5 percent of GNP. Parks have traditionally been a large part of that industry. Much of the early momentum for establishing reserves was stimulated by tourist-oriented commercial interests, particularly the railroads. When the hot springs at Banff received legislative protection in 1887, Prime Minister Sir John A. Macdonald announced: "The intention is to frame such regulations as will make the springs a respectable resort, as well as an attractive one in all respects."[5]

Macdonald's intentions shaped the purpose of management policies for years to come. A 1938 Parks Bureau publication, for example, reads like an advertisement: "Among Canada's greatest tourist attractions are her National Parks, areas of outstanding scenic beauty or interest which have been set aside for the use and enjoyment of people."[6] Occasional warnings of overcrowding were heard: "What is happening in the United States' parks is an indication of what can be expected in the Canadian parks."[7] Nevertheless, parks, services, and roads leading into parks were increasingly developed to facilitate tourist conveniences.

TOWN SITES.   The most apparent manifestation of providing conveniences to park visitors are the town sites that are located in western parks. The town around the hot springs at Banff actually predates the park. When the federal government did take over the land at Banff—as well as at Jasper, Prince Albert, Riding Mountain, Waterton Lakes, and Yoho—townspeople were not removed. Instead, the government leased their land and property back to them on long-term (usually twenty-one-year or forty-two-year) contracts with virtually perpetual renewal. Until the 1960s the government used low taxes and other practices to encourage the development of towns and businesses around these original leaseholders. Not surprisingly, the town sites grew into permanent settlements.

During the 1960s and 1970s park managers occasionally disagreed with townspeople over specific developments or projects, but for the most part, town sites were accepted as a fact of Canadian park life. Any systematic attempts to change policies, with higher lease rates or tougher renewal provisions, inspired claims of deprived citizen and counterclaims of provincial rights. Court decisions generally preserved the status quo.

RECREATION.   Most Canadians were blessed with miles of open spaces already near their homes, so many early visitors to the parks sought something more than scenery. Park authorities, concerned with attracting and pleasing visitors, responded to this desire by allowing forms of recreation that some say violated the pristine nature of parks. Sport hunting and commercial fishing were allowed and even encouraged in many park areas. Wardens attempting to maintain sizable game populations shot predators on sight until 1937. Motorized boating and golfing were popular activities at many parks. Later, park managers offered downhill skiing and snow-mobiling to turn their sites into year-round attractions. Canadian author-ities even offered to use the parklands surrounding Lake Louise in Banff for the Winter Olympics in 1968 and 1972.

Emphasis on providing recreation stems, of course, from the "enjoy-ment" mission stated in the agency's mandate. This language was used to justify any commercial activities providing recreation in national parks. In fact, the agency's attitude toward recreational activities was much stronger than merely allowing them. "In practice, recreation, often backed by large commercial interests, dominated national parks."[8]

RESOURCE EXTRACTION.   National parks were viewed as useful for eco-nomic activities other than tourism and recreation. Logging, coal mining, and oil drilling have occurred inside and outside the parks for decades. In the earliest years of the Canadian parks, resource exploitation was a focus of parks policy. In fact, during this period, "the clash was not between preservation and exploitation, but between exploitation for tourism and that for minerals and timber."[9] Even these uses did not always clash. For example, an early coal mine near Banff was viewed as a tourist attraction.

While internal encouragement of these activities was reduced in 1911 and again in 1930, pressures to allow resource extraction accelerated out-side the parks over the decades. Between 1878 and 1930, economic policies were based on the exploitation of Canada's natural resources. Pressure from the logging industry grew as Canada became the foremost exporter of forest products in the world, contributing, for example, almost one-third of the globe's newsprint. In more recent decades, the nation has consciously pursued energy self-sufficiency in oil and gas supplies. These developments shaped and pressured the management of parks for decades.

Parks as an Afterthought

Management of Canadian national parks was delegated to the National and Historic Parks Branch of the Department of Indian Affairs and North-ern Development. The agency developed a structure similar to that of the U.S. National Park Service, decentralized in implementation responsibil-ities but susceptible to political influence. Decentralized implementation was a logical response to the remoteness and uniqueness of individual Canadian park areas. The 1911 act itself granted broad powers to the park wardens, positions comparable to those of American rangers. Nevertheless, from the Macdonald policy of 1887, political authorities encouraged the parks agencies to allow the use, rather than the preservation, of these wild spaces.

This political intervention into agency management of parks was somewhat counter to Canadian traditions of political deference to administrative authority. In this policy area, however, politicians showed little respect for the managers of the national parks through the first three-quarters of the twentieth century. This lack of deferential behavior resulted from traditional Canadian perceptions of wilderness areas and from the role of the provincial governments.

TRADITIONAL PERCEPTIONS. One major reason the Canadian parks agency received so little respect was that these public lands were not historically perceived as precious. Faced with an abundance of wild, road-less areas, most Canadians were not that interested in other wilderness areas that were set aside as somehow special. One parks bureau director admitted in 1970 that "parks to Canadians come pretty much as an after-thought."[10] The preservation aspect of the park mandate seemed somehow a luxury when so much of the country remained in relatively natural condition. For example, even during the environmental revival of 1970, a public opinion poll showed an even split on whether or not people wanted more paved roads in the parks.

Nevertheless, the parks themselves tempted political consideration. Placed within a cabinet department that was often seen as inefficient, if not inept, the parks, as Jasper's superintendent Gaby Fortin told me, were "always the good news." Politicians thus showed the managing agencies little respect but willingly intervened in specific park affairs when they needed some good news. Good news often translated into finding new ways to use or to develop the parklands.

PROVINCIAL GOVERNMENTS. The economic interests encouraging development and use of the parks found particularly receptive ears at the provincial level of Canadian government. The willingness of politicians to intercede in national park management on behalf of economic interests was facilitated by the important position of provincial governments. Most areas of environmental protection have historically been dominated by the provinces.

The importance of provincial authorities is certainly true in the issue of public lands. The Canadian constitution itself does not provide many simple answers to federal-provincial questions, particularly on questions of natural resources. The original British North America Act of 1867 favored a strong central government, and it established residual authority clearly with the federal government, unlike its American counterpart. The provinces did retain ownership, however, of natural resources, and they strengthened their claims on the use of public lands when the constitution was amended in 1982. As a result, parklands are consistently subject to the tension between the central government and the ten provincial governments. The interactions between federal and provincial authorities can be critical to the management of national parks.

Provincial governments have traditionally held an important role in national parks. Since 1930 new national parks have been created only when a provincial government obtains the rights to the land and then transfers those rights to the federal government. Provincial governments were reluctant to relinquish control of any territory to the federal government if the land held potential for resource development. As a result provincial parks, more accessible to development than their national counterparts, flourished. Provincial parks comprise nearly 60 million acres of land, twice that of the Canadian national parks and seven times that of American state parks. Even when land has been turned over to the federal government, however, provincial politicians are necessarily concerned with the economic health of their provinces and have thus developed close ties with park managers in the use of their lands for tourism and resource extraction. Often national parklands are bordered by provincial lands. As the assistant deputy minister of the department wrote in 1979, "provinces generally seek a clear commitment that benefits will accrue to the provincial and local economy in the short term. This has led to large-scale capital developments in and near many new national park units within the first five years of park establishment."[11] Thus park managers were subjected to political influence from both central and provincial authorities even after the parks were established.

The Canadian parks agency entered the 1970s in a condition similar to that of the NPS. The agency was housed in a multifaceted cabinet department, faced with serious demands for development, subject to political pressures from all levels of government, and decentralized in its organizational structure.

The Most Turbulent Decade

The Canadian national park system was shaken by developments between the mid-1960s and mid-1970s. By the end of that period, many Canadians recognized that change was essential for the national park system. At the ten-year reunion of the 1968 Conference on Parks, Assistant Deputy Minister A. T. Davidson looked back and said: "We have just come through the most turbulent decade in the history of Canada's national parks."[12] Later observers seconded this assessment. The major develop-

TABLE 3-1. *Visits to Rocky Mountain Parks, Selected Years, 1950–67*
Thousands

Years	Banff	Jasper	Kootenay	Yoho
1950–51	450	86	97	51
1955–56	701	160	289	n.a.
1960–61	1,077	370	524	72
1966–67	2,044	630	773	925

Source: Nelson (1970b, p. 64).

ments of the period included increased use of the parks in existence, expansion of the system, growing environmental awareness, and the creation of regional offices.

INCREASED USE.   One major development was the perceived overuse of the few existing parks. While few national parks were added between 1930 and 1970, access to the ones that already existed was dramatically improved. For example, the road from Edmonton to Jasper opened in 1931, and the scenic Banff-Jasper highway was accessible by 1940. Correspondingly, visitor services and accommodations expanded in the town sites of Banff and Jasper. No longer dependent on the railroads and enticed by town site attractions, visitors flocked to the Rocky Mountain parks. Table 3-1 shows the increase in visitors to four of these parks in representative years. The steady increase in visitors is particularly dramatic in the 1960s. This growth was not limited to these mountain parks. While the Canadian population grew between 1956 and 1966 by only 25 percent, the number of visits to national parks rose by 222 percent.

EXPANSION.   Increased visitation stimulated public demand for more parks, but park system management remained virtually static. The 1968 Conference on Canadian Parks concluded that the system needed to expand. As one analyst noted, too much of "the brunt of this [visitor] increase is falling on relatively few parks."[13] Indeed, while fourteen parks existed in 1930, only six new ones had been established by 1968. Between the mid-1960s and the mid-1970s, ten new national parks were created, more than doubling the land area under protection to 50,000 square miles. Such significant expansion was made possible by public demand, supportive political authorities, a healthy economy, and a new focus on unsettled northern lands.

ENVIRONMENTAL AWARENESS.   The late 1960s and early 1970s were a period of increased environmental awareness throughout North America. Perhaps the signal for this reawakening in Canada occurred in 1966. The

Canadian government and the province of Alberta, having lost the competition to host the 1968 Winter Olympics by a single vote, reapplied for the 1972 games. In early 1966 the Canadian bid, promising considerable development of parklands in Banff, seemed to be the front-runner, ahead of bids from Salt Lake City and Sapporo, Japan. In the interim before the International Sports Federation decided the 1972 host areas, however, the Canadian Wildlife Federation expressed concern about the effects of proposed developments on birds and wildlife in Banff. The contract was awarded to Sapporo, and one member said Banff lost because the Olympics "had enough troubles without being bothered by bird-watchers."[14] Albertan authorities and others were disappointed about losing the opportunity to attract tourists and foreign income, but the event was an early manifestation of the impact of environmentalism on Canadian parks.

Subsequent developments illustrated the growth of Canadian environmentalism. The Twenty-sixth Parliament, meeting from 1968 to 1972, passed nine major environmental statutes, including the Clean Air and Clean Water acts. The cabinet-level Department of the Environment was created in 1971. Interest groups also grew. While they were not nearly as organized as their U.S. counterparts, Canadian environmentalists did form the National and Provincial Parks Association in 1963, a Sierra Club branch in 1970, and the Canadian Nature Federation in 1971. Greenpeace formed shortly thereafter and became one of the prominent groups in Canada. Environmentalism became a major social issue during this period.

Public interest in more natural forms of recreation, specifically related to national parks, grew dramatically during the 1970s. For example, between 1967 and 1976, the percentage of Canadian hikers quadrupled and of canoers tripled. Increased awareness of national parks and wilderness areas fostered reconsideration of how those areas were managed. Numerous recommendations were made throughout the decade for greater emphasis on research, better developed master plans for the parks, more realistic fees, and increased use of zoning for various areas in the parks.

REGIONAL OFFICES.   These growing demands on the parks had structural effects. By the end of the 1970s the Canadian park system was an unsettled mix of central planning and decentralized implementation. Political intervention remained high, but now public input into park management was also a fact of life. Central offices were increasingly involved in system expansion and planning, but the rapid growth of the system necessitated a continued reliance on the people at the field level for specific decisions. One major response to these developments was the establishment of regional offices. Five regional offices were established in Halifax, Quebec, Cornwall, Winnipeg, and Calgary. The regional offices were established to integrate central policies and field-level programs and to improve relations with provincial governments. The regional offices quickly exerted their authority. When each region began to use different zoning systems for categories of parklands, for example, the central office was forced to step in to ensure that "five national parks' systems were not to evolve."[15] By 1979 management of park activities was thus shared between levels of government with no fundamental agreement on apportionment of responsibility.

The Agency at the End of the 1970s

Even with all of the changes in the preceding years, the managing agency of the Canadian national parks retained its reputation for emphasizing economic and political uses well into the 1970s. At the least, the agency and its forerunners did little to distinguish themselves from the early resource-use attitude of many of the Canadian citizens and their government. The reputation had its roots in the focus of original parks established around town sites and railroad destinations but was nurtured by the "historical policy relics" described earlier. The agency thus reflected "a pioneer mentality of 'unlimited' forests, lakes and wildlife" that precluded emphasis on preserving natural conditions.[16] By the late 1970s scholars and officials within the agency were encouraging attention to the preservation goal.

Changes in the Political Environment

Between 1979 and the present, two major changes in the political environment of the Canadian parks agency significantly affected the management of the national parks. Increased political awareness of environmental concerns among the electorate led to reconsideration of priorities among agency goals, and the agency was given a new structural home and a significant increase in political support.

Sensitivity to Environmental Issues

In June, 1985, Minister of the Environment Suzanne Blais-Grenier announced that any new national parks might have to come with compromises, in effect allowances for logging and mining. Two months and a substantial political furor later, she was gone, replaced by Tom McMillan, who announced when he took office that he would not allow logging or mining in the parks. The incident typified an increased political awareness of public concern for environmental issues in general and the protection of national parks in particular. John Carruthers, a parks agency official described the new political attitude toward parks as "much more sensitized to broader environmental issues." That philosophy translated into a restatement of the goals of the agency. Heightened sensitivity resulted from increased threats, growing environmental awareness, and the receptivity of political leaders to environmental demands.

THREATS.   Yoho National Park, in eastern British Columbia, has beautiful mountain scenery, clear-flowing streams, undeveloped back country, and one major road. Unfortunately for the wildlife in the park, that road is Route 1, the Trans-Canada Highway, the major east-west artery across the country. Route 1 bisects the park, following the Canadian Pacific railroad corridor. Thus, even though the park hosts only about 700,000 tourists and 6,000 overnight visitors annually, more than 3 million drive through every year. Much of the traffic consists of eighteen-wheel trucks that do serious damage to the road. Wildlife road kills are too numerous to count. The CPS spends a great amount of time and resources cleaning up the road kills and repairing the highway. Superintendent Ian Church calls the Trans-Canada the major threat to natural conditions in Yoho. The extent of this problem for Yoho and other Canadian parks is unparalleled in the United States.

The CPS published its own *State of the Parks* report in 1990, ten years after the NPS version. While similar in scope and purpose to the NPS document, the CPS description of threats to the parks differed in a subtle but illustrative manner. Rather than presenting threat categories with counts of affected parks, the CPS consciously lists each park unit separately with descriptions of its own threats. As the document argues, "overall condition ratings for an entire system are not the best indicators for the conditions of individual resources."[17] The document profiles more than 150 units. The concentration on individual parks rather than on the system as a whole is an early tip-off to the decentralized structure. Otherwise, many of the threats facing Canadian parks are quite similar, with exceptions like the highway in Yoho, to those affecting U.S. parks to the south.

The most obvious threat to the Canadian parks, as to the U.S. parks, is visitor impact. What U.S. park managers call "surges" in tourists, Canadians call "peaking."[18] Tourist traffic peaks in the summer months after school is out, particularly on weekends. The mountain parks, especially, are swamped by residents from Calgary, Edmonton, and Vancouver as well as by foreign visitors. The lines, overcrowding, and noise that result make aesthetic appreciation a challenge. The perfect illustration is a drive along the remarkable Icefields Parkway between Jasper and Banff national parks. On busy summer days, picnic areas are jammed, tour buses pollute the view, and driving often resembles a stock car race through the mountains. Table 3-2 provides the simplest measure of visitation in num-

TABLE 3-2. *Total Visits to CPS Sites, 1989–92*
Millions

Units	1989–90	1990–91	1991–92
National parks	35.85	35.59	36.72
National historic sites	8.26	8.43	8.70
Historic canals	.88	.82	.67
National total	44.99	44.84	46.09

Source: Canadian Parks Service (1992g, pp. 1–22).

bers of person-entries a year. As the table shows, visitors to Canadian national parks alone totaled nearly 37 million in 1991–92.

The effects of tourism became increasingly apparent in the 1980s. Increased awareness resulted in part from the sheer growth in tourist numbers. Because the Socio-Economic Branch of the CPS is currently refining its measures of visitor use to reflect more sophisticated concepts, obtaining consistent measures over time is difficult. Still, since the parks hosted 18.5 million in 1977–78, the figure in table 3-2 shows a 100 percent increase over the preceding fifteen years. Even while the current total is somewhat less than the U.S. total of 58 million, the visitors are spread over fewer total acres (46 million compared with 56 million acres), and the total represents a much higher percent of the Canadian population than does the 58 million of the U.S. population.

Beyond aesthetics, the infrastructure suffered from these higher levels of use. Tourists scrambling for camera angles erode delicate areas, and off-road travel destroys winter range for animals. Controversies at individual parks over the impact of tourism included such issues as wildlife protection, landscaping, alcohol consumption, and possible use of quotas. As one conference concluded, "Tourism may ultimately destroy not only the natural values, but itself."[19] Even tourist trade associations noticed the higher levels and explicitly acknowledged that "a healthy tourism industry requires a quality environment."[20]

The increased highway traffic resulting from tourists and other users threatened not only Yoho; other parks bisected by the Trans-Canada Highway include Banff, Glacier, and Revelstoke. Jasper is traversed by the Yellowhead. Since Route 1 and the Yellowhead are the major east-west routes across the Canadian Rockies, the Yoho highway problem is replicated in these other parks. In 1983 alone, for example, 120 big-game animals were killed by cars in Banff. This relatively concentrated passage creates pressures on western Canadian parks unlike any in the United States. Figure 1-1 showed that potential use of the most popular Canadian parks is heavily concentrated by population location and by accessible highways. Again, even though population levels and visitor totals in Canada are lower than in the United States, nowhere in the U.S. park system is the potential tension between human economic activity and natural preservation resulting from highways so inherently intense.

Recreation demands also threatened the natural condition of the parks in the 1970s and 1980s. The most visible controversy involved downhill skiing in the Rocky Mountain parks. Four ski resorts exist in Banff and Jasper: Sunshine, Norquay, Lake Louise, and Marmott. These resorts were built with the encouragement of the government, which was eager to find year-round uses for the parks. Attempts to expand Lake Louise in the 1970s and Sunshine in the 1980s attracted considerable attention. The proposal to expand the upper village at Lake Louise to provide accommodations, clubs, and theaters for 3,000 overnight visitors inspired an intense debate that resulted in a compromise low-growth option of visitor services but no permanent upper village. Public awareness of this issue was reflected by the filing of more than 2,000 briefs of opposing viewpoints on the expansion proposal. Advocates of the 1989 expansion of Sunshine promised 200 permanent new jobs, $28 million for the local economy, and no effect on the environment. These promises inspired support from the Alberta provincial government, but skepticism and opposition from environmentalists.

Recreation controversies were not limited to expansion of these two ski resorts. Debates, reminiscent of the Olympics controversy, over allowing competitive events on park ski facilities raged in 1986. Proposed ski resorts also aroused opposition in Prince Albert and Riding Mountain national parks. Other controversial recreation issues have involved golf courses, snowmobiling, and mountain biking.

Many of the people who visit or work in the western parks consider expanding the town sites to provide recreational and other opportunities a threat to the natural conditions of the parks. Pressures for growth at Banff are infamous. As the town expands, interactions with wildlife multiply and intensify. The CPS identifies the major internal threat to Banff as "pressure to develop montane valley bottom lands for visitor use vs. protection for wildlife."[21] The CPS must also monitor town sites in Jasper, Kootenay, Prince Albert, Revelstoke, Riding Mountain, Waterton Lakes, and Yoho.

Another potential complication to managing the parks has even deeper historical roots than town sites—property claims of indigenous peoples. Many existing and proposed new national parks contain land either previously or currently occupied by indigenous Indian and Eskimo tribes. The 1978 Parks Conference predicted: "The relationships between the activities of indigenous people and the restrictions traditionally associated with national parks and other reserves is a major problem for the next ten years."[22] Indeed, some controversies have arisen. Predictably, they often involve attempts by indigenous peoples to prevent transfer of land to parks. The expansion of Bruce Peninsula in 1989 is an example.

Another source of issues are the rights of indigenous people to maintain traditional ways of life on federal lands. Whether or not traditional hunting and fishing pose a threat to the parks is an interesting question, but regardless of the answer, resolution poses a challenge for the CPS.

Internal "historical policy relics" of resource extraction affected parks well into the 1980s. Current boundary lines on several parks have been redrawn to allow provincial development of lands. Waterton Lakes was cut in half to allow logging and mineral exploration. Prince Albert sacrificed nearly 1,000 square kilometers to logging and farming interests. The government allowed resource extraction activities to continue within some park boundaries rather than reducing the size of the park. The most controversial of these cases is in Wood Buffalo, where loggers have clear-cut hundreds of square kilometers.[23] Having renewed a long-term lease in 1982, Canadian Forest Products, Limited, planned to continue to cut the last white spruce forests in Alberta until 2002. While the CPS encouraged the company to use improved cutting practices, environmental groups led by the Canadian Parks and Wilderness Society took the case to court. Many feared the timber would be gone by the time the case was decided, but the CPS may have speeded resolution by announcing in 1992 that continued logging would be a violation of CPS regulations. Federal spokespersons admitted to reporters that they were embarrassed by the whole situation: "We've inherited something that we're not comfortable with."[24]

As in U.S. national parks, protection of Canada's national parks traditionally stopped at park borders, but developments outside the parks have significant implications for parklands. An article in *Canadian Geographic* concluded about the parks: "They might be considered as reasonably respectable oases in a desert of environmental outrage."[25] External threats are becoming increasingly critical to the future of Canada's parks. It should be pointed out that these threats are not generated exclusively outside of park boundaries, but rather constitute developments that cross the permeable membranes that demarcate public lands.

Because of past commercial activities, resources lying just outside park boundaries are often being used. Another analyst used an analogy comparable to the "oases" remark: "Most of Canada's 35 parks have now become islands in seas of unbridled development."[26] Riding Mountain is encircled by agricultural development. Grasslands has been permeated by cattle grazing and oil exploration. More than 75 percent of the borders at La Mauricie have been clear-cut. Timber companies have demanded that park boundaries of Pacific Rim be changed to facilitate logging. Oil wells have been sunk on the land that was previously part of Waterton Lakes. Wood Buffalo, already damaged by logging practices, is home to the Peace

River, polluted from paper mills and drying up from hydroelectric power development. Finally, Yoho is surrounded by mining and logging operations as well as the booming resort town of Golden. When I asked Superintendent Church about the effect of these threats on his park, his low-key response spoke for many parks: "The impacts are major."

Wildlife populations have always been affected by sport hunting and commercial fishing both inside and outside of park borders. Obviously, animals do not recognize political demarcations, and since Elk Island may be the only park whose boundaries are fenced, they are thus in danger immediately after walking, flying, or swimming over park lines. Each fall roughly 30 percent of La Mauricie's moose population is killed outside the park. Riding Mountain, home to the biggest black bears in the world, loses several of them each year to hunters using baiting points just outside the park. This hunting is legal and even endorsed by the Manitoba Provincial wildlife branch as maintenance of a "sustainable yield."[27]

Animal populations are also affected by illegal hunting, especially poaching. Again, these activities were abetted by past government policies, which used such low fines for poaching—for example $500 for bighorn sheep—that many saw them as "just a license to hunt."[28] For years CPS wardens and officials were frustrated by the low fines. Finally their efforts were rewarded with 1988 amendments that made Canada's poaching fines among the highest in the world. Still, poaching is difficult to stop. Yoho superintendent Church described his park's efforts to track grizzly bears beginning in 1987. After three years, park rangers had put collars on eleven bears, but only three could be found. The others had been victims of poaching or road kills. As Church said, "We were losing bears faster than we could get collars on them." The public has become much more aware of the poaching issue largely as a result of efforts by the CPS itself to attract support.

Wildlife populations can also be affected by disease. The tragic story of the Wood Buffalo bison herd has generated a great deal of attention recently. Again, the story has its roots in past management practices. In the 1920s, 6,000 plains bison infected with cattle brucellosis were shipped to Wood Buffalo, where, because of negligence, they bred with 1,500 wood bison. Now, with cattle herds nearby risking infection, the park is under considerable pressure from Agriculture Canada, ranchers, and some biologists to kill the entire herd. Others say the risks are exaggerated. Nevertheless, the situation poses a highly visible challenge for the CPS.

Finally, air pollution represents a growing external threat. Canada's national parks are just as affected as other areas of the country by the acidified precipitation produced by coal-burning power plants in western provinces and in the midwestern United States. Acid rain ruins property, damages crops, and kills aquatic life. According to the comprehensive

Green Plan document, eight out of ten Canadians live in areas with high levels of acid rain. Pukaskwa, located on the north shore of Lake Superior, is particularly susceptible to acidified conditions in its lakes and streams. La Mauricie, in Quebec province, is severely affected by acid rain, to the point where the "natural integrity and complex balances of its ecosystems are at considerable risk."[29]

Public recognition of the acid rain problem has intensified along with growing political involvement in the issue. The Mulroney government announced plans in 1985 to cut contributing sulfur and nitrogen oxide emissions in half over the next nine years. Even then, the government admitted that progress ultimately depended on corresponding efforts from the United States, which Canadians have been demanding for years. The 1990 U.S. Clean Air Act provided some response in mandated reductions of sulfur dioxide emissions, particularly from plants in the Midwest upwind of Canada and in the northeastern United States. Title 4 of the legislation calls specifically for cooperation with Canada in these efforts. A subsequent pact between the two nations in 1991 furthered that process. The ultimate effect on Pukaskwa, La Mauricie, and other Canadian parks is not yet clear.

ENVIRONMENTAL CONCERN.   In the 1970s and 1980s, the environmental awareness of the late 1960s grew into a general public concern. Increased awareness of threats to the parks and public support for environmental concerns became influential factors in setting goals for the national parks.

Public opinion was important regarding the environment generally and parks specifically. The salience of environmental issues actually declined somewhat in the early 1970s. When asked, however, the public expressed a high level of receptivity to environmental concerns. In 1979, one survey showed that 89 percent considered environmental deterioration a major concern, and two-thirds of those were more concerned than they had been five years earlier. Another survey in 1983 reiterated the strength of these feelings, particularly in regard to the parks. Finally, a nationwide Environment Canada survey in 1987–88 found broad levels of support for the parks. Roughly 26 percent of respondents were frequent users of parks, with another 43 percent "favourably disposed" toward them. The 25 percent not disposed toward visiting a park were generally older and farther away from possible park destinations.

Interest groups had less influence on parks policy than in the United States, but they were still an increasingly important factor. Canadian interest groups have traditionally not been as organized or politically sophisticated as American groups. Furthermore, environmental groups were generally excluded from political decisionmaking processes and were assumed to be represented by government actors. In addition, many Canadian environmental groups traditionally concentrated on issues other than those exclusively affecting the parks, such as reducing acid rain from the United States. Nevertheless, recent years have witnessed the greater inclusion of interest groups, including environmentalists, in political bargaining procedures. Some groups have recently emulated American models, particularly the Canadian Parks and Wilderness Society (formerly the National and Provincial Parks Association), which is a counterpart to the National Parks and Conservation Association (NPCA) in the United States.

The national groups have been joined at all levels of the park system. As one authority relates, "Each province experienced a burst in the growth of local groups."[30] Further, most of the parks have their own support organizations, which have various titles and contribute to the parks in different ways. For example, members of the Friends of Prince Albert Park run the bookstore and hold special events on park grounds. Other specific groups include such diverse organizations as the Federation of Ontario Naturalists, the Western Canada Wilderness Committee, and the Society of the Trail of the Great Bear.

The activities of these groups are as diverse as the groups themselves. Several organizations are well enough established at the national level to criticize general policy statements and directions. These national groups, particularly the Canadian Nature Federation (CNF) and the Canadian Parks and Wilderness Society, participated, for example, in removing Blais-Grenier from office. Canada's chapter of the World Wildlife Fund and the CNF collected more than 350,000 signatures to demand expansion of the park system. Others have testified in much more specific events, such as the operation of the Banff airport. The Canadian Parks and Wilderness Society took government officials to court over the logging in Wood Buffalo. These organizations also enhance public attention to park activities by attending conferences such as the World Congress on Parks, the most recent of which attracted 1,500 participants from more than 130 countries to Venezuela in 1992. Finally, they engage in activities to educate the public and to attract future members.

RECEPTIVITY OF POLITICAL LEADERS.   Canada's top political leaders from the late 1970s to the present were receptive to the growing environmental concerns in society. Four different administrations governed the country during the period, although the governments headed by Joe Clark and John Turner were too short-lived to have much effect on the management of the parks. The governments of Pierre Trudeau and of Brian Mulroney expressed support, in varying degrees, for the sentiment behind emphasizing preservation in park management. Once past the rhetoric, however, both showed only mixed commitment to reversing the traditional emphasis of using the parks for economic growth.

The two Trudeau governments, separated briefly by Clark's term in 1979, advocated reform of government policies and preached sensitivity to the natural and human environments. The early Trudeau years witnessed liberal attempts to use big government for a variety of causes. For example, in 1975, the Trudeau government attempted what one witness called "the most exhaustive, most encyclopaedic, and most determined attempt ever made by any government any time, anyplace, anywhere, to develop a logical, systematic, rational program of policy priorities."[31] Overall, Trudeau's government displayed a strong willingness to use the central government to develop priorities and to systematize programs. Trudeau had committed himself to protecting natural reserves and his Liberal party to preserving the "sanctity of the national parks from commercialism."[32] Further, Trudeau and Minister of the Environment Jean Chrétien followed up on promises to greatly expand the parks system. In fact, Chrétien made a bet in 1968 with an environmentalist that he would add ten new parks to the system. The five-dollar bill is now framed in his office.

Trudeau also had other priorities in mind, however. He believed in an efficient use of resources to provide for economic growth. The administration advocated national energy policies and the development of natural resources, particularly oil, to strive for energy self-sufficiency. In attempting to meet demands from energy producers, the provinces, and competing interest groups, Trudeau often settled for only incremental changes. When pressed on difficult issues facing the national parks, Trudeau and his associates refused to fall on the sword of preservation. For example, Chrétien backed off from early attempts to charge market rates for leases in the western town sites. Later, in the fight over the Lake Louise ski resort, the Trudeau government refused to take a controversial position on expansion after Alberta authorities waffled on support for the development. Ultimately, the Trudeau governments provided a receptive ear for preservation demands but did not provide the commitment necessary to ensure a change in the way national parks were managed.

When Mulroney's Conservative party took power in 1984, the partisan shift produced certain expectations. One Conservative member of Parliament saw it as a green light for tourism: "We should encourage thousands of these international people from other countries, to come and enjoy our scenery. Such a policy will bring new dollars and help our balance of trade as well as provide for jobs for people."[33]

Mulroney, like Ronald Reagan, was supported by pro-business and pro-development constituencies. Nevertheless, Mulroney's own agenda on most issues, let alone parks, was not clear. In addition, Mulroney soon became heavily involved in constitutional matters involving Quebec and national unity, which diverted his attention from other issues such as parks.

The Mulroney government was well aware of the political costs involved in threatening environmental values. The Blais-Grenier case is illustrative. Blais-Grenier dismantled the Canadian Wildlife Service, eliminated several research programs through budget cuts, refused to meet with senior bureaucrats or environmental leaders, and attracted negative publicity for taking an April-in-Paris fling on public money. Her efforts stalled, however, when she tried to alter the mission of the national parks. She made some remarks encouraging logging and mining, which sparked immediate criticisms from environmental groups and editorial writers and a denunciation from Mulroney himself. Mulroney pointedly muzzled the controversial minister and publicly rejected her statements on parks. Shortly thereafter, the prime minister backed his verbal reaction by replacing Blais-Grenier with Tom McMillan. The new Environment minister, who had said "parks policy is tourism policy" while serving as Tourism minister, now said that he would err on the side of preservation. The administration's rapid damage control provided a sharp contrast to Reagan's toleration of Interior secretary James Watt's numerous provocations.

Mulroney's management of the national parks, however, like Trudeau's, evolved into supportive rhetoric without consistent commitment. Rhetoric on national parks provided a means to respond to environmental demands without antagonizing provincial authorities or regional economic interests. By the time Mulroney took office, the legitimacy of provincial involvement in natural resource policies was already recognized. As the decade progressed, regional authorities became more powerful and their support became more valuable in the fight over national unity. Mulroney displayed an early awareness of the need for concessions to western regional interests by repealing Trudeau's despised national energy program. The Mulroney government continued to attempt to avoid antagonizing provincial authorities by not obstructing efforts to expand regional tourism and development. Thus, when situations arose involving national parks and provincial interests, the Mulroney government kept a low profile. Like Chrétien during the Lake Louise controversy earlier, for example, Environment minister McMillan refused to take a position on the Sunshine ski resort expansion in 1989.

Ultimately, Mulroney proved himself "a political animal . . . who would do whatever was necessary to stay in power."[34] On environmental issues in general, the Mulroney government offered "limited, relatively symbolic responses [that] were sufficient to maintain legitimacy for the system."[35] In the area of national parks policy, this meant verbally acknowledging the value of preservation without drastically altering CPS behavior.

RESTATEMENT OF FEDERAL GOALS.   Management goals for the CPS were officially clarified in policy documents and in the *Green Plan*. The 1988 amendments and the 1991 policy statement for the agency not only reiterated the original mandate but clarified CPS priorities by calling for the encouragement of "public understanding, appreciation, and enjoyment of this natural heritage so as to leave it unimpaired for future generations."[36] The epitome of the Canadian government's rhetorical commitment to environmental values was called the *Green Plan*. The Mulroney government prepared the *Green Plan* after hearing thousands of ideas and recommendations at hundreds of public meetings on environmental concerns throughout the country. Released in 1990, the *Green Plan* contained more than 400 recommendations for environmental actions, outlines for 100 initiatives, and commitments for $3 billion in addition to the $1.3 billion already spent by the government on environmentally related programs. The *Green Plan* called specifically for emphasizing preservation and establishing an explicit timetable for system expansion.

For all of the attributes and innovative ideas contained within these documents, they are, after all, only paper. Without serious commitment of material resources and programs, these directives could be criticized as naive at best, deceptive at worst. Some critics contend that the monetary figures contained within the *Green Plan* were a sleight of hand, simply a replacement of funds that had already been cut or moved elsewhere. Bill Waiser, author of a fine book on Prince Albert National Park, called the document "a sham."[37] Others felt that none of the Mulroney rhetoric, including the commitment to environmental sensitivity, was either realistic or sincere.

Nevertheless, the rhetorical commitment was potentially important. The *Green Plan* rhetoric was reinforced by numerous other policy statements and environmental assessments. Conservation ministries were established at both federal and provincial levels. The National Task Force on Environment and Economy recommended greater inclusion of environmental interests in future public deliberations at federal and provincial levels. Together these developments established a verbal commitment to certain ecological ideals, including refocusing park management on the original mandate for the agency, "mountains without handrails." That commitment legitimized and sanctioned the real changes taking place within the CPS.

Let the Managers Manage

Political support for the CPS has increased since 1979 in important ways. While monetary resources remained flat, the agency received a new institutional home and an increase in autonomy. The CPS experience thus illustrates the rather counterintuitive insight that agencies can welcome and enjoy greater levels of political support even without greater financial commitment.

THE BUDGET SQUEEZE.   Ken, a retired Air Force officer, shoved another log onto the campfire. The temperature was dropping nearly as fast as the sun over Lake Audy in Riding Mountain National Park. The cool night air provided a welcome replacement for the retiring dragonflies as our protection against southern Manitoba's early summer mosquitoes. I had joined Ken and his family around the fire for conversation, marshmallows, and a little less-benign evening refreshment. "We come up here every summer," Ken was saying. "We always used to go to Lake Katherine, but the park did not have the money to keep it open this year." I realized that their changed travel plans were my good fortune, but also that they represented something more serious. Because the CPS had not received a budget increase, facilities such as the Lake Katherine campground had to be closed.

The Canadian park system experienced serious budget cuts in the 1980s, like the NPS during the same period. The financial squeeze began in the late 1970s. Although outlays to the parks continued to grow through the 1970s, they did not keep up with the rapid expansion of the system. Existing funds were spread thinly as the creation of new parks drained money from the older units. The Mulroney government made further reductions explicit. Environment Canada's budget was cut by $33 million in 1984 alone. Environment minister Blais-Grenier celebrated the park system centennial with a variety of budget cuts and staff eliminations. Even though her tenure as Environment minister did not last long, fiscal and economic conditions ensured that the budget would continue to be curtailed. Indeed, different regions and parks used a variety of means, such as campground closures and early retirement arrangements, to cut spending levels in the 1990s. For example, Banff National Park was forced to retire 7 percent of its staff in 1991. As the president of the Canadian Parks and Wilderness Society responded, "The parks service has been getting the economic squeeze for a while, and it seems a very strange place to make cuts."[38]

On the revenue side of the equation, two other issues were also similar to U.S. experiences. First, entrance and user fees in Canadian parks are almost identical to those in the United States. The annual pass, for example, costs $25 in the United States and $30 in the slightly less valuable Canadian currency. Just as with the U.S. fee revenues, park revenues go back into the general fund. Canadian park fees were raised in 1975 and 1985 for reasons similar to those expressed in the U.S. debates. For instance, the government stated explicitly in 1984 that it wanted to cut the fiscal deficit by, in part, charging more for services such as park management.

A second aspect of revenues points out an important difference between the two park services. Whereas individual U.S. national parks use a single concessioner, Canadian parks host numerous homes and businesses, particularly within the town sites. Revenue is collected in Canada, not as a percentage of sales, but as lease payments by individual leaseholders. By the start of the 1980s, the parks held more than 5,000 leases. Lease rates have historically been incredibly low. Some Banff leaseholders, for example, were paying as little as $30 a year in 1979. The parks bureau sets the rates, and in the early 1980s it occasionally raised rates as much as 10 percent a year. Despite encouragement from environmentalists to raise lease rates even more, however, these increases have inspired serious opposition. Finally, a recent development concerning leases is oddly similar to Interior secretary Manuel Lujan's anti-Matsushita campaign over the Yosemite concessioner issue. For years the Japanese have been buying up and taking over leases in Jasper and in Banff. In 1989 Banff park authorities, at the behest of Alberta politicians, began imposing restrictions on Japanese ownership in the park. Nevertheless, regardless of who owns the leases, the CPS has not received a revenue windfall from either higher entrance fees or more expensive leases.

ENVIRONMENT CANADA.   The institutional structure of the national parks agency was changed significantly in 1979. The 1979 Government Organization Act and associated orders-in-council moved the national parks agency, Parks Canada, from the Department of Indian Affairs and Northern Development into the recently created department, Environment Canada. The move followed years of increasing public concern for the environment and pressure to give Parks Canada more autonomy. For example, a spokesman for the watchdog group National and Provincial Parks Association called the agency's former home "so multifaceted that the interest of national parks was not best served by being placed there."[39] Jasper superintendent Fortin, who witnessed the switch from the field, described how park programs were often compromised by other departmental operations and then stated succinctly: "The department was always screwing up." Advocates argued that the reorganization responded to environmental demands and also made sense for the parks agency's former home. At least according to some, the department could now focus its efforts on Indian activities and concerns.

While some parks advocacy groups had pushed for a separate parks department, Environment Canada was not an unwelcome alternative. Environment Canada had been formed out of various components within the

FIGURE 3-1. *Formal Structure of Environment Canada, 1990*

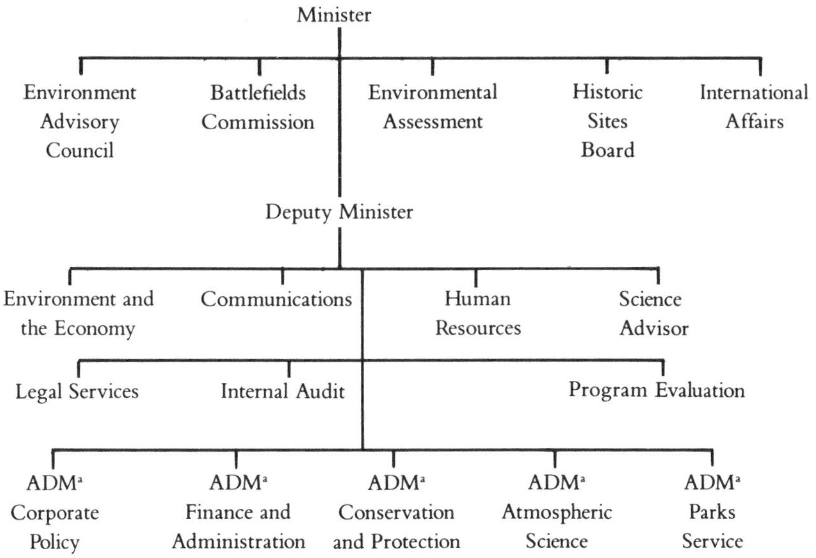

Minister

| Environment Advisory Council | Battlefields Commission | Environmental Assessment | Historic Sites Board | International Affairs |

Deputy Minister

| Environment and the Economy | Communications | Human Resources | Science Advisor |

| Legal Services | Internal Audit | Program Evaluation |

| ADM[a] Corporate Policy | ADM[a] Finance and Administration | ADM[a] Conservation and Protection | ADM[a] Atmospheric Science | ADM[a] Parks Service |

Source: Canadian Chamber of Commerce (1990, p. 203).
a. Assistant Deputy Minister.

government that dealt with environmental issues. The organizational structure of Environment Canada, shown in figure 3-1, is clearly focused on environmental concerns. Interest groups supporting parks thus accepted the move as a means of facilitating better coordination, research, planning, and management. As strategic policy analyst John Carruthers told me, the move shifted the agency's broader focus from land planning and social issues to environmental concerns such as acid rain and the relevant role for the parks.

Largely as a result of the move into Environment Canada, the level and degree of interagency disputes affecting the CPS in the 1980s were much less severe than those involving the NPS during the Reagan-Bush era. As figure 3-1 shows, Environment Canada, unlike the U.S. Department of the Interior, focuses exclusively on environmental concerns. The department does not house agencies supporting dam-building or strip-mining, as does Interior. In general, Environment Canada has matured since its establishment to become more innovative and more inclusive of environmental interests during decisionmaking procedures. Specifically in the area of parks, top officials in the CPS called the difference between former and current placement for the agency noticeable. Several officials described the agency as "more integrated" with the rest of the department.

In addition, political support for the CPS was enhanced by the absence of competing agencies, whether in an Interior-type department or not, dealing with public lands at the federal level. The structure of forest management, in particular, is quite different in Canada from that in the United States. While NPS disputes with the U.S. Forest Service are often contentious and publicized, CPS dealings with forest lands are more localized and quiet. CPS disputes over forests, particularly on lands bordering the parks, are quite common. Except for Wood Buffalo, however, the negotiations do not occur on national levels. The reason is simply that forest management in Canada mainly resides with provincial ministries of forestry. As a result, environmental groups do not have easy national targets like John Crowell, Reagan's head of the U.S. Forest Service, and disputes receive less national attention. Recently a proposal to place the parks under forest jurisdiction received some attention, but it was immediately rejected.

FREEDOM IN THE FIELDS.   While the move into Environment Canada provided some insularity for the CPS specifically, the shift toward providing more autonomy for bureaucratic agencies was consistent with the trend throughout Canadian government. The lack of deference to authority was replaced by bureaucratic empowerment, as individual agencies gained "considerable freedom in their fields of jurisdiction."[40] The Canadian bureaucracy now enjoys "much more" discretion than the American.[41]

The effort to enhance discretionary authority culminated with the release of *Public Service 2000* at the end of the 1980s. One manager after another in the CPS summarized the document's meaning to me as, "Let the managers manage." This broad statement of bureaucratic autonomy is not, at least according to the CPS employees, just more paperwork. All those I talked to agreed with the sentiment expressed by Pukaskwa superintendent Mike Murphy: "I've heard this stuff before, but for the first time, action is backing up the words."

Like its American counterpart, the Canadian bureaucracy continued to be increasingly involved in nearly all aspects of Canadian life. Unlike the United States, however, decisionmaking within government, fueled by the demand for expertise and rational planning, has increasingly shifted from politicians to bureaucrats. This bureaucratic growth in Canada, unlike the United States, has not automatically inspired negative reactions from the public but rather has occasionally fostered a "more sympathetic attitude" toward bureaucratic behavior.[42]

Changes in the Operating Environment

As a warden for more than twenty years in Banff and Kootenay, Perry Jacobson saw the changes in the CPS from the ground up. When I asked him why they had occurred, he answered: "This organization did some real soul-searching." The answer is important not just in its explanatory power, but also in its emphasis on the agency. The changes in the CPS came from within.

Changes in the political environment—rhetorical emphasis on environmental concerns and increased autonomy for bureaucratic agencies—facilitated changes in the operating environment for the CPS. Direct political intervention in agency affairs diminished. Employees of the CPS took the concept of empowerment to heart and decentralized much of the agency's decisionmaking structure. Largely as a result of the first two conditions, morale in the agency climbed.

Focus on Policy

As a public agency, the CPS is always accountable to the political process. Indeed, Parliament and the prime minister's cabinet enjoy many tools for influencing agency actions. Important differences from the U.S. system exist, but the potential to alter agency behavior is evident. As Superintendent Church told me after describing his version of growing agency autonomy, "We are still a political beast." The political pressure on the CPS has been much less than than the micromanagement experienced by the NPS in the 1980s, however. As Bob Speller, a member of Parliament told me, politicians are "not that involved except for broad policy."

AUTHORIZING LEGISLATION.   Parliament uses some tools similar to those used by the U.S. Congress in its relations with public agencies. Most obviously, Parliament creates public agencies and establishes their mandates and their responsibilities. In the context of the parks, field managers, just like their U.S. counterparts, consistently say their decision process begins with authorizing legislation. CPS planning documents explicitly state that the first step in any process is to review legislation. Members of Parliament can thus amend the agency mandate or alter the emphasis within mission statements.

One important difference in this aspect of potential legislative involvement is that while each U.S. park is created under a separate legislative act, all Canadian national parks are established under the authority of the National Parks Act of 1930. For example, legislation creating a national park on the Mingan Archipelago in 1984 stated: "The National Parks Act applies to the lands set aside as a reserve . . . as if the reserve were a park within the meaning of that Act."[43] This unifying act potentially precludes some of the nuances seen in U.S. park creation, such as exemptions to specific regulations.

CASE WORK AND MONITORING.   Members of Parliament (MPs) represent individual districts (or ridings), as do members of Congress, and they engage in case work and monitoring. They may thus become involved specifically in the actions of individual parks within their jurisdiction. Jasper superintendent Fortin described the MP from his region, Joe Clark, as having been "fighting with the park for the last twenty years." As Fortin admitted, occasionally Clark has forced the park authorities to take actions, such as permitting the Jasper airport, that they would not otherwise have pursued. Other managers, sounding remarkably similar to U.S. park superintendents, said they communicate with their MPs on a regular basis. In one sense, the relationship between representatives and parks in their ridings could be more intense in Canada. Because of the presence of town sites, an MP has permanent constituents within the parks, rather than just visitors, as in the United States. This difference should not be exaggerated, however. Even though town sites do not exist inside U.S. parks, the parks share their representatives with the countless towns that have sprung up just outside park borders.

Backbench members of Parliament do not possess the same tools as do individual U.S. legislators, however. Even though party discipline may be relaxing somewhat to facilitate more independent behavior by MPs than in the past, the situation bears little resemblance to that in the United States. Individual MPs cannot direct appropriations to pork-barrel sites, as Representative McDade did with Steamtown, unless they are ministers, except in very special circumstances. Furthermore, because party discipline is so tight, members of the opposition party have difficulty affecting parks in their ridings. For example, Yoho superintendent Church said his district's opposition-party MP "can't get involved too much and he wouldn't get what he wanted anyway." As a result, many parks are relatively insulated from pressure from MPs.

Monitoring is more difficult for MPs because they do not possess the staff and resources of their U.S. counterparts. Even while power and resources spread out somewhat through the Canadian system in the 1980s, the executive-centered nature of the parliamentary system developed in the early years has continued through the present.[44] The lack of resources is particularly important in legislative attempts to monitor agency behavior. MPs lack the information necessary to control the actions of field-level bureaucrats. One trail-level bureaucrat, a park superintendent, was quite explicit. He had just explained to me all the information that he possessed, which was unavailable to his MP, even if the MP had wanted it. "I can control him," the bureaucrat said, with emphasis that indicated he knew it was contrary to the usual perception that politicians control bureaucrats. It was also a line unlike any I had heard in the United States. In fact, it illustrated a broader point made by one of Canada's leading parliamentary scholars: "A recent study has discovered that senior public servants in Canada tend to relegate MPs to the periphery of the policy-making world more so than those in the United States."[45]

Members of Parliament do interact, as mentioned above, with CPS field managers and officials. The interaction is generally much more supportive and informative than that described in the Jasper airport situation. Even Superintendent Fortin, who detailed that case, asserts that MPs are much more supportive now than they were in the past. He attributes the improvement partly to the greater prestige the CPS has now compared with its junior status before the move into Environment Canada in 1979. Superintendent Mac Estabrooks of Riding Mountain refers to his relations with MPs as "contact, not interference." In case after case, Estabrooks said that the MPs supported him 100 percent. Elk Island chief warden Jack Willman admitted that involvement may vary from park to park or even region to region but added that he was "not aware" of any excessive intervention. Others suggested that low-profile parks might receive less attention, but even those in the most visible parks cited low levels of political intervention. Banff superintendent Charlie Zinkan called political involvement in Canadian parks "not close to the level" of that in U.S. parks. Virtually every warden and superintendent I interviewed in Canada characterized their relations with politicians as anything but micromanagement.

On the other end of the parliament-bureaucracy relationship, MPs consistently told me they did not get involved in the details of park affairs. Many MPs do not have park units in their ridings. Others who do also denied intense interaction with field managers. One told me his last confrontation with the park in his riding was seven years ago. Party discipline does affect this relationship. After all, as Conservative MP Ken James told me, the head of Environment Canada was a fellow party member. Thus if MPs wanted to intervene in CPS behavior, they would be forced either to question their own party or, recalling Church's characterization of his representative's opposition status, to complain in vain. Neither makes much electoral sense. As MP Speller said when I asked whether they intervened in park affairs, "Why would we?"

ADMINISTRATIVE CONTROLS.   The CPS, as part of Environment Canada, is subordinate ultimately to the prime minister through the minister of the environment. The administration can affect the parks through personnel changes, budgets, directives, and policy guidelines without even passing legislation. As a specific example, Riding Mountain superintendent Dave Dalman discussed how a ministerial signature is necessary to rezone park areas in the management plan to allow different uses. Evidence that intense political pressure from the administration is possible was apparent in the expectations that the government of Prime Minister Mulroney, largely personified in Environment minister Blais-Grenier, would have a severe effect on the parks. Pukaskwa superintendent Murphy recalled fearing that "Many in the Mulroney group wanted to dismantle us."

Had the Mulroney government stuck to what many perceived as its original intentions, the characterization of low political intervention might be hollow. The reasons for low administrative intervention into the parks are largely related to political changes. The Conservatives were committed to administrative delegation and at least willing to respond rhetorically to demands for environmental responsiveness. Mulroney's guiding beacon, as nearly all sources agree, was pragmatism. His behavior in the Blais-Grenier case and in pushing the Reagan administration on acid rain controls indicated that he was aware of the strong concern for the environment among the electorate. This awareness was noticed at the field level as well. Several trail-level bureaucrats recounted watching the Mulroney group back off on environmental issues early in the term.

Finally, political intervention in the CPS remained low because it was not stimulated, as so often happens to the NPS, by powerful interest group demands. Again, neither environmental nor economic groups are as capable of pushing political intervention into agency actions as are their counterparts to the south. "Canadian decision makers enjoy more autonomy from interest groups than their counterparts in the U.S. On the one hand, they are more insulated from pressure by environmental groups, which may make them less sensitive to environmental concerns. But on the other hand, they are also more insulated from business pressures, and may thus be free to represent environmental interests, if they wish. Thus, in the Canadian system, the policy interests of decision makers themselves may be more important."[46] With little interest group and political intervention, CPS employees assumed responsibility for the decisions affecting national parks.

Taking It to the Field

In 1990 the Western Region office of the CPS published its plan for the future, *Choosing Our Destiny*. The title, in more than just a symbolic sense, tells the story of this chapter. The CPS, as a result of the changes in the political environment emphasizing ecological concern and bureaucratic autonomy, was not predetermined to a particular destiny but rather was able to choose its own structure and behavior. *Public Service 2000* gives a hint as to the likely structural direction of the agency. If current political sentiment favored delegation, why not continue to delegate all the way down to the field? Indeed, the CPS, particularly in the Western Region, has decentralized authority and responsibilities down to the trail level in recent years.

WHEN?   Pinpointing structural change is not always possible. While an agency could suddenly decide to decentralize, the more gradual evolution of the CPS is probably more typical. Internal decentralization of the CPS started in the mid-1980s and accelerated in the last part of the decade. Prince Albert superintendent Dalman, involved in various planning sessions, puts the starting date at around 1984 or 1985. At least as early as 1986, officers in the Western Region were using the term "decentralization" to describe changes in decisionmaking procedures.[47] Banff superintendent Zinkan concurred on the timing, saying the changes have been developing since the mid-1980s. Jasper superintendent Fortin said the effects at the field level had been noticeable for at least three years. The structural changes accelerated during the late 1980s and early 1990s, culminating in a dramatic reorganization of the Western Region on May 11, 1992.

Why did change start at that time? Political authorities were receptive to and, in fact, supportive of efforts to empower line workers. The Mulroney government encouraged reliance on middle managers and field-level employees. While structural changes did not occur uniformly across the board, the Conservatives' rhetoric and documents such as *Public Service 2000* created an atmosphere conducive to decentralization efforts. As one analyst observed after four years of Mulroney, "there has been an increased emphasis given to administrative decentralization within many departments."[48]

Specific to the policy area of parks, the CPS had been part of its new department long enough to become confident of external support for alterations. Further, increasing demands on the agency combined with static budgets made it necessary to explore new ways of managing. Environmental concerns and park use were growing simultaneously. New parks were perceived as underfunded and undermanaged. Legislation such as the pending amendments to the National Parks Act, which passed in 1988, was creating new demands on the agency. Finally, people inside and outside the CPS were viewing the agency as increasingly stagnant, bloated, and inefficient. As Jillian Roulet, an official of the Western Region, recalls, "We were losing respect."

WHERE?   The answer to the question of where change developed is definite. As I traveled through Canada, CPS employees from Ottawa to British Columbia pointed to the Western Region as the example to be followed.

Even those in the neighboring Prairie and Northern regions who are proud of their own recent evolution describe the Western Region as "about a year ahead" in development. Several officials outside the west used the term "dramatic" to describe the recent activity in that region.

The Western Region is the pioneer in these changes for two reasons. First, the Western Region, with headquarters in Calgary, contains the most famous, most historic, and most visited units in the Canadian system. Units in the Western Region host roughly two-thirds of the annual visits made to Canadian parks. Correspondingly, the region generates about 65 percent of the total CPS revenue. As a result, employees within the region recognize that they are, according to the regional plan, "in a unique position to respond to both threats and opportunities."[49]

A second reason is the personnel of the Western Region. Regional headquarters received an infusion of new leadership in the late 1980s, particularly in the person of Sandra Davis. Davis, a former management consultant, and others were hired from outside the CPS to bring private enterprise perspectives to the agency. While some managers were uncomfortable with these new perspectives, other regional officers such as Roulet and David Bowes adopted them as "welcome" changes. Trail-level bureaucrats in both the Prairie and Western regions referred to Davis and her assistants (like Roulet) as "dynamic" and "capable," providing a vivid contrast with the field-level views of many superiors that pervade the NPS.

HOW?   Mac Estabrooks, a veteran of thirty-five years with the CPS, is now superintendent of Riding Mountain. When I asked him if the CPS had changed in recent years, he responded shortly: "Yes, much." When I followed up by asking how, he replied simply: "More empowerment at lower levels." Estabrooks's view was restated and expanded time after time, and in stronger and stronger terms, as I traveled toward the Western Region. In the Rocky Mountains, Kootenay's Jacobson noted cogently: "This time, they really are taking it to the field."

The CPS is an agency in transition. The changes are so recent that outside documentation is difficult to find. Nevertheless, the structural changes started in the Western Region have been adopted throughout the organization. At the system level, decentralization is apparent in, first, the independent behavior of individual regions. The regional offices, themselves fairly recent creations, tend to operate somewhat independently of each other. This tendency, stimulated by greater responsibilities, became particularly evident recently. Regional responsibilities remained relatively static until the 1980s, Roulet told me; then more and more responsibilities were delegated to the regional level. The evolution has continued until now, when, as the Prairie Region's Richard Leonard says, "the regions are so separated that each is carving out its own future." The strategic plan of

FIGURE 3-2. *Formal Structure of the CPS, 1991*

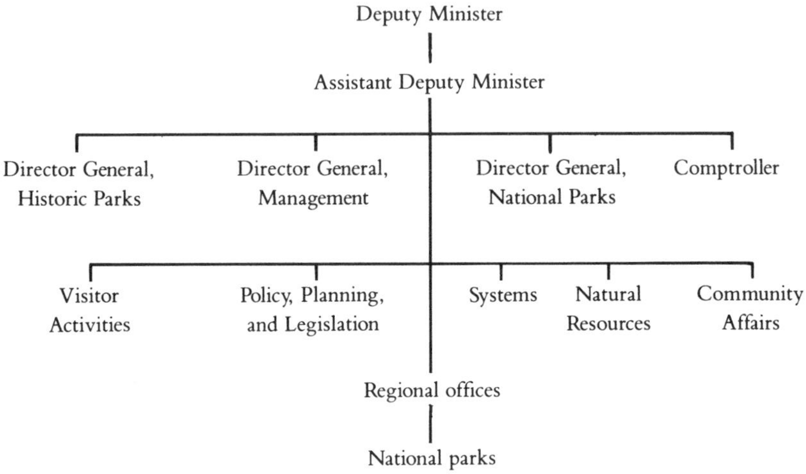

Deputy Minister

|

Assistant Deputy Minister

Director General, Historic Parks	Director General, Management		Director General, National Parks	Comptroller
Visitor Activities	Policy, Planning, and Legislation	Systems	Natural Resources	Community Affairs

Regional offices

|

National parks

Source: Unpublished CPS document, 1991.

each region is presented in documents such as *Choosing Our Destiny* for the Western Region and *A Bridge to the Future* for the Prairie Region. While the similarities between the plans are numerous and important, the documents are prepared autonomously to reflect the specifics of each individual region.

Second, systemwide decentralization has been endorsed by the central offices of the agency and by Environment Canada. Divergent regional growth is not only acknowledged by central officers of the CPS, but also applauded. Donna Petrachenko, director of Strategic Policy and Program Affairs, argued that diversification only makes sense considering the different problems and challenges faced by each region. Environment Canada, the cabinet-level department that houses the CPS, published a document on the transition of the entire department in 1991. A major principle stated in the document is "to enable managers to manage" through such steps as empowerment, delegation, and reduction of management layers.[50]

Third, systemwide decentralization, while not readily apparent in the formal organization chart for the agency, is evident in the allocation of resources. Figure 3-2 shows a formal organization chart for the structure of the CPS at the start of the 1990s. Table 3-3 displays the allocation of money and personnel to different activities of the agency. The category of operations captures the resources that are spent directly on the parks. As the table shows, more than 73 percent of expenditures and 79 percent of person-years are used at the parks level. These totals provide a stark contrast to the figures presented in chapter 2 for the centralized NPS.

TABLE 3-3. *CPS Expenditures for Fiscal Year 1991*
Percent of budget

Activity	Cost	Person-years
Park operations	73.3	79.1
Park development	10.3	7.8
Management/administration	16.4	13.1
Total	100.0	100.0

Source: Environment Canada (1992, pp. 4-27, 4-40, 4-50).

TABLE 3-4. *Percent of CPS Expenditures by Central Offices*

Activity	1986–87	1992–93
Park operations	3.6	2.6
Park development	31.4	28.8
Technical services	37.7	39.4

Sources: Environment Canada (1987, p. 4-18) and Environment Canada (1992, p. 4-15).

How much change over recent years does this represent? Ten years ago, Parks Canada spent 64 percent of its budget on operations. Table 3-4 gives a more detailed breakdown of how the agency has decentralized in recent years. The figures show the percentage of money being spent within park service activities by central offices as opposed to expenditures in the field. The percentages show small but important changes. Resource expenditures in operations and development have decentralized, but more resources are currently being spent by the central office in technical service. The agency has decentralized its management capability to the field level but retained the central ability to analyze and to provide technical support.

Structural change has also occurred at the regional level. Regional offices, even while operating independently, have all chosen to go the route of decentralization. Most used a similar process to make that decision. The process is built around the participation of field managers in all important decisions. One means for accomplishing that participation is through regional management boards. These boards draw together regional directors with superintendents and other relevant personnel to develop consensus management. The story at the start of this chapter illustrates how, even though such meetings also took place in the past, field input is now essential to the process. In the Prairie Region, for example, the board usually meets twice a year for a week at a time. The meetings decide important policy issues but also provide an opportunity, as Prince Albert superintendent Dalman says, "to get the field and the offices on the same wavelength."

Related to the board meetings are ad hoc task forces created within the regions. These task forces address specific issues or programs of limited

FIGURE 3-3. *Formal Structure of Western Region, 1992*

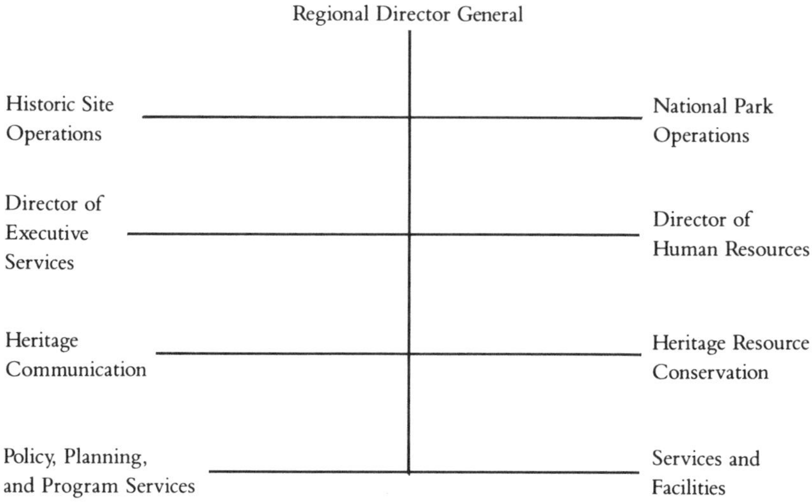

Regional Director General

Historic Site Operations		National Park Operations
Director of Executive Services		Director of Human Resources
Heritage Communication		Heritage Resource Conservation
Policy, Planning, and Program Services		Services and Facilities

Source: Unpublished CPS document, 1992.

duration; their goal is to produce recommendations for the entire region. In 1989, for instance, the Western Region instituted a series of task forces to provide direction for all CPS activities. The task forces were staffed by people from the field as well as the regional office, and they were often chaired by superintendents. The Integration Task Force, chaired by Superintendent Roger Beardmore of Revelstoke-Glacier examined means to coordinate CPS behavior with Environment Canada activities. The Science and Protection Task Force, to cite another example, was chaired by Charlie Zinkan, now superintendent of Banff. Zinkan cited these task forces as one means for field personnel to be fully involved in regional decisions.

A third and more painful way that regions have decentralized is by eliminating management personnel. Figure 3-3 shows the structure of the Western Region. Some categories, such as the resource management specialists in the Heritage Resource Conservation division, reflect an emphasis that is different from that of the NPS. The formal charts do not show the changes taking place in regional offices. The cuts in the Western Region following the May 11, 1992, reorganization were substantial. As Jillian Roulet, one of the survivors, said, the changes required personnel reductions that "created a lot of uncertainty." Superintendent Dalman described it more graphically: "The big bang on May 11 disemboweled the old organization and caused substantial human stress in the Western Region." While the cuts in other regions have not been as severe yet, CPS employees

FIGURE 3-4. *Formal Structure of a Western Region Park, 1991*

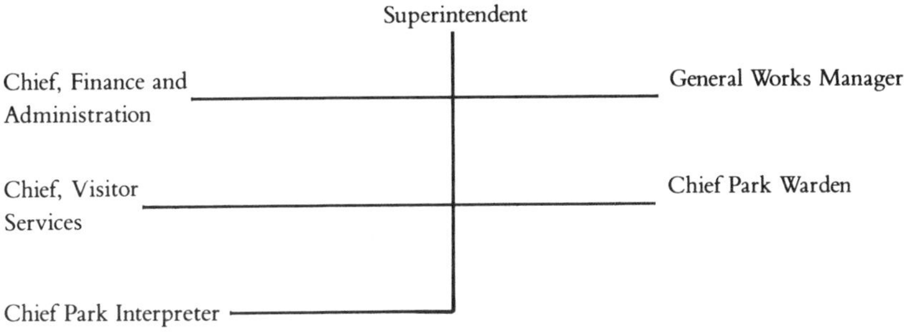

Superintendent

Chief, Finance and
Administration

General Works Manager

Chief, Visitor
Services

Chief Park Warden

Chief Park Interpreter

Source: Unpublished CPS document, 1991.

like Dalman certainly noticed it and are well aware that other regions have often emulated Western developments.

Fourth, a recent development indicates that other regional offices are accelerating the decentralization of their structures. Both the Atlantic and the Prairie regions are establishing district offices in remote areas to further delegate regional responsibilities. For example, the Prairie and Northern Region is now establishing district offices for the East Arctic and the West Arctic to take over decisionmaking responsibilities in the north.

Finally, decentralization is noticeable especially at the park level. Again, these changes are most evident in the Western Region. Developments there reflect both the emphasis on empowerment and the orientation toward business principles of delegation and accountability.

Because the changes in the CPS are coming from lower levels, the formal organization charts of individual park units are revealing. Figures 3-4 and 3-5 show typical organization structures of parks in the Western Region before and after the May 11 changes. Several changes are apparent. First, whereas nearly all planning used to be done at the regional level, now each park has its own chief of Planning. This change reflects an obvious delegation of responsibility down to the field level. Second, visitor services and interpretation are now combined into Heritage Communication. This not only streamlines the organization but affects policies. Third, each park has added a chief of Technical Services to update research and analysis for the individual unit.

Other Western Region changes are less obvious. Staffing classifications previously made at the regional level are now park-level responsibilities. In facilities management, whereas parks used to have a General Works Manager in charge of all facilities, now western parks have a Services and Facilities department built around product lines. Typical product lines for a park may include front country, scenic corridors, and campgrounds. If a

FIGURE 3-5. *Formal Structure of a Western Region Park, 1992*

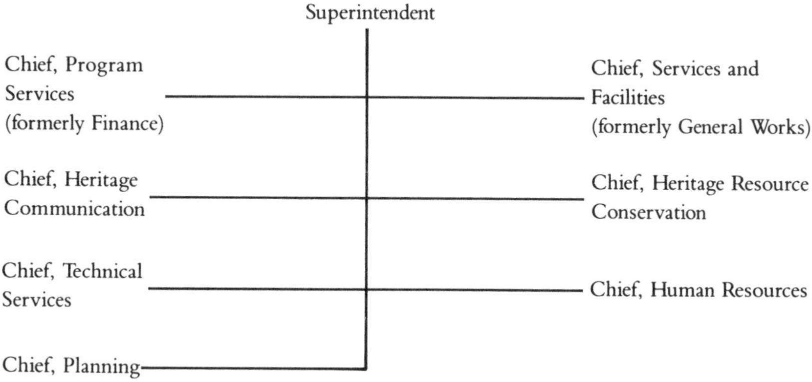

Superintendent

Chief, Program
Services
(formerly Finance)

Chief, Services and
Facilities
(formerly General Works)

Chief, Heritage
Communication

Chief, Heritage Resource
Conservation

Chief, Technical
Services

Chief, Human Resources

Chief, Planning

Source: Unpublished CPS document, 1992.

problem occurs in a campground outhouse, for example, the plumber will report to the person in charge of campgrounds rather than directly to the General Works Manager. This change induces accountability down to the level of the individual employee.

At parks throughout the system, expenditures are increasingly decided at the field level. In general, as Kootenay's Jacobson describes, "They give you a lump sum and you spend it on what you need." Changes in two specific expenditure categories illustrate Jacobson's larger point. First, parks previously were allotted a certain number of person-years for specific jobs and had little flexibility in allocating them. Now field managers can take the money for those salaries and spend it as they see fit. As Pukaskwa superintendent Murphy says, "That's a big change." Second, field managers now have new discretion in the expenditure of capital expansion and development funds. Previously parks would submit specific project information proposals to the regional office, which would then divide up whatever pool of capital funds was available. Now, as Elk Island's Willman puts it, "they give us a target figure and it's up to the park how to spend it." Decisions up to $30,000 are made entirely within the park. As regional officer Roulet says, "We don't even want to see capital projects up to $30,000." This freedom facilitates much more long-range planning at the park level than was possible before the changes. It also provides an obvious contrast to the NPS, where all design projects initiated in the field must be sent to regional or higher levels.

Officials throughout the CPS agree on the overall decentralization that these changes have accomplished. The field managers in the Western Region unanimously acknowledge that authority has shifted to their level.

Roulet summarizes the Western Region's approach to field managers: "Now we give them a framework and then turn them loose." In the Prairie Region, Prince Albert superintendent Dalman says of his immediate superiors, "Now they set broad objectives and within that is a lot of latitude." Prairie regional officials concur that superintendents now have "much more discretion." In the Ontario Region, Pukaskwa superintendent Murphy sees his regional office in "more of an advisory than a control capacity" and calls it a "refreshing change." In Ottawa, CPS official Carruthers says simply, "We want to leave as much as possible to the parks."

WHY?   Why did the CPS decentralize? The structural tendencies of an autonomous agency, as organization theorists point out, are unpredictable. They may centralize, decentralize, or do neither. Traditional theory predicts strong centralizing pressures: "the power to make decisions and to interpret and complete the rules, as well as the power to change the rules or to institute new ones will tend to grow farther and farther away from the field where these rules will be carried out."[51] Yet employees in the CPS chose decentralization for the structural form. Why?

Several reasons are readily apparent. Countless personnel cited the *Public Service 2000* document as a call for empowerment. The Mulroney government was encouraging delegation and decentralization. Yoho's Church called Mulroney's views on the parks simply consistent with what was going on "throughout the world." Many emphasized that the "new blood" in the organization, particularly in the Western Region, had injected a businesslike philosophy of delegation and accountability.

Certainly all of these factors contributed, but none alone is convincing. Documents like *Public Service 2000* are often ignored. The Reagan administration preached the same decentralization rhetoric as the Mulroney government, without comparable effect. Furthermore, not all agencies in Canada under Mulroney did decentralize. Mulroney centralized his own office by creating the office of deputy prime minister. The Office of Inspector General of Banks enlarged its staff from forty-five to seventy following the controversies of 1986. New blood and new ideas, like paperwork, are often avoided. As Elk Island's Willman says, "We're normally pretty entrenched; we don't like a lot of change."

Nevertheless, the agency decentralized. The most important factor to remember was that change came from within. Officials at the field and regional levels initiated and nurtured the changes in their own organization. Why did field-level employees push decentralization? CPS employees pushed for decentralization because they think decisions made in the field will be "better" decisions. The system was expanding and facing new challenges that required local expertise. As Jasper's Fortin said, "The big bosses in Ottawa have never worked in the parks, so they have to rely on the people in the field."

Happily Paid in Sunsets

Pukaskwa superintendent Murphy has been with the CPS for more than twenty years. He also worked with the U.S. NPS at earlier duty stations. When I asked him to compare the personnel of the two agencies, he referred to recent publicity about low morale in the NPS and said, "You just won't find that up here." Murphy described the personnel in both agencies as dedicated and committed, but the difference is that fear of downsizing is the only major source of displeasure among Canadian employees, whereas U.S. rangers face that threat as well as dissatisfaction with agency behavior. Other CPS employees echoed Murphy's comments.

In most ways the field employees of the two agencies are comparable. In Canada they are called wardens rather than rangers. But their jobs, their pay levels, their living conditions, and even their uniforms are quite comparable. Both have proud traditions; the Warden Service dates back to a force of "former trappers, mountain men, and even reformed poachers" in the early 1900s.[52] Most of the superintendents in both services rose up through the ranks. Perhaps more important than any other aspect, their compensation similarly involves being "paid in sunsets."

While the attitudes of employees of the two services toward the parks may be consistent, opinions toward recent changes in their agencies are not. Riding Mountain superintendent Estabrooks even used the term "family" I heard in the United States when he compared the two systems: "We're more dedicated to our family than they are down there."

Comparison of morale levels between the two agencies cannot be quantified, but the relatively high level cited in Canada is an almost inevitable result of the increased autonomy and decentralization. CPS employees perceive independence in their decisionmaking and feel that their actions have consequences. Other studies have shown how morale can be affected by organization structure. The lack of empowerment caused by rigid central planning can stifle initiative and cooperation. Central emphasis on formalized rules can weaken incentives for dedicated field workers. One study concludes, "The people who work for government will always want the freedom to do their work without excessive constraints."[53]

Stories about abysmal living conditions or dissension, in contrast to the U.S. situation, are conspicuous by their absence. Lacking material about dissension or low levels of morale, the news stories on Canadian park employees generally center on what the wardens see as threats to the parks themselves. One warden, for example, wrote an editorial entitled, "It's time to realize national parks can't be all things to all people."[54]

Another long-time warden, Sid Marty, has written articles and books that describe potential dangers but are mostly positive toward the agency. Former superintendent Steve Kun has said critical things about the town site at Banff and about past policies, but his criticism has not been directed at the current CPS management. Nearly all of the wardens and other field employees I interviewed in Canada expressed a similarly positive attitude toward their agency and its future.

Changes in Policy Behavior

How have these recent changes in the CPS affected policy behavior? Given the green light to pursue an ecological focus and the room in which to operate, CPS employees have decentralized responsibilities and decisionmaking authority down to trail-level bureaucrats who accelerated shifts in the agency's policy emphasis in the 1980s and early 1990s. In short, the CPS is now more likely to manage parks to preserve natural resource conditions than to provide political and economic usefulness.

Created to Represent

Unlike Steamtown, a visitor to Grasslands National Park in southern Saskatchewan rarely sees hard hats or even ranger hats. The money that is available for the park is not being used for construction but rather to buy up grazing rights from local ranchers. One of the newest CPS units, Grasslands obviously was established for reasons other than economic and political. Grasslands is one of many park units added to the CPS system in the 1980s to represent a different ecosystem. The consistent effort by the CPS to preserve different natural regions of the country provides sharp contrast to the NPS experience with sites such as Steamtown.

CRITERIA.   The criteria for inclusion of an area in Canada's park system are similar to those stated for the NPS. Agency policy identifies parks as "areas which are representative of the major natural environments of Canada." Historic sites are to "commemorate a historic event of national importance or preserve any historic landmark or any object of historic, prehistoric or scientific interest of national importance."[55]

In its 1991 *State of the Parks* report, the CPS listed managed sites in the categories shown in table 3-5. The table shows categories that are somewhat similar to those in the United States, but without the individual nuances that create so many distinctions in the U.S. system. All parks in Canada are created under the same legislative authority, and all historic sites are established under part II of the National Parks Act. This pre-

TABLE 3-5. *Summary of Units of the CPS, 1990*

Type	Number
Canals	7
National historic sites	102
National marine parks	2
National parks	27
National park reserves	7
Total	145

Source: Canadian Parks Service (1990e, various pages).

scribed uniformity precludes some of the extra categories used in the United States. National park reserves, such as Nahanni and Kluane, are differentiated from national parks in that land claims by indigenous people are respected to allow traditional hunting, trapping, and fishing. Also, the number of historic sites in Canada is currently more than 700; the number given in table 3-5 represents those actively managed by the CPS.

New park units are supposed to be established under a systematic approach to expansion. Unrepresented regions or themes are studied to identify worthy areas. Areas should represent the natural features of the region or the historic theme and contain minimal human impact. In order to select one area, candidates within a single region are compared with regard to factors such as exceptional biogeographic features, accessibility, recreation opportunities, and socioeconomic issues such as potential costs. The process is designed to recognize traditional uses of the land, to incorporate provincial demands, to use cost-sharing between levels of government for acquisition, and to foster public involvement in decisionmaking. Once an area is selected, a new unit proposal is prepared to determine feasibility and public support. Then negotiations are conducted to transfer the land or property from the current holders, generally provincial or territorial governments, to federal authority. Finally, legislation establishes the new park, reserve, or historic site. The process can take years or even decades.

REALITY.   The Canadian experience in creating new parks differs from the U.S. experience in that efforts to put theories of preserving diverse ecosystems into practice have met with some success. In 1979, the CPS identified nine marine and thirty-nine terrestrial "natural regions based on physiographical, biological, oceanographical, physical, and geographic parameters."[56] These regions are shown in figure 3-6. An additional 100 potential areas were identified and proposed for study. At that time, the system represented only 40 percent of its potential, while several regions,

TABLE 3-6. *Priority Historic Themes for Representation*

Short-term	Long-term
Fishing	Public works
Basque whaling	Mining
Ranching	Energy development
Immigration	Commerce
Forest products	Manufacturing
Prairie settlement	Canada and the world
Native history and northern	Engineering and architectural
native commemoration	achievement

Source: Canadian Parks Service (1990e, p. 65).

such as the Rocky Mountains, were densely represented with numerous popular parks.

The progress in the past fifteen years has been impressive. Since the call for expansion at the 1978 conference, eight new parks or park reserves have been formally added to the system. All but one, Bruce Peninsula, represent regions not previously included in the Canadian system, and Bruce surrounds Canada's first national marine park. The marine park, Fathom Five, was an immediate result of approval of the National Marine Parks Policy in 1986. Of the eighteen terrestrial regions still not represented, areas have been selected in fourteen, and active proposals exist in seven of those. The system plan was later expanded to include a priority for historic themes. Table 3-6 shows the plan for filling in representation of historic themes in the parks system. Numerous historic sites have been established or are currently in the final stages of negotiation. This progress has been applauded and at least rhetorically supported by Environment ministers Tom McMillan, Lucien Bouchard, Robert deCotret, and Jean Charest throughout this period of time.

Grasslands is the antithesis of Steamtown in the United States. As the park's own brochure says, "Grasslands National Park is being created to represent undisturbed mixed grass prairie. There is not now nor likely to be another such park in any park system in North America." The area was selected to represent region thirteen in the system plan. It was certainly not picked to deliver political pork to the region. In fact, the CPS has only been able to proceed slowly on completing the park because it must acquire the land from reluctant cattle ranchers. Even the CPS admits only "qualified" local support for the Grasslands park idea. The initial agreement between Canada and host province Saskatchewan was signed in 1981. For the next five years, terms concerning oil and gas exploration and water resource management were negotiated. The final agreement of 1988 prohibits oil and gas development in the park. The preservation emphasis is

FIGURE 3-6. *Status of National Park Natural Regions*

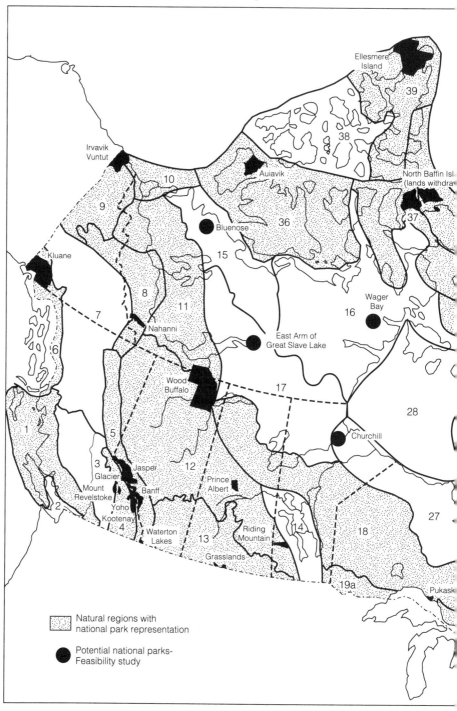

Source: Canadian Parks Service, unpublished, August 1993.

WESTERN MOUNTAINS

1. Pacific Coast Mountains
2. Strait of Georgia Lowlands
3. Interior Dry Plateau
4. Columbia Mountains
5. Rocky Mountains
6. Northern Coast Mountains
7. Northern Interior Plateaus and Mountains
8. Mackenzie Mountains
9. Northern Yukon Region

INTERIOR PLAINS

10. Mackenzie Delta
11. Northern Boreal Plains
12. Southern Boreal Plains and Plateaus
13. Prairie Grasslands
14. Manitoba Lowlands

CANADIAN SHIELD

15. Tundra Hills
16. Central Tundra Region
17. Northwestern Boreal Uplands
18. Central Boreal Uplands
19. (a) West Great Lakes · St. Lawrence Precambrian Region
    (b) Central Great Lakes · St. Lawrence Precambrian Region
    (c) East Great Lakes · St. Lawrence Precambrian Region
20. Laurentian Boreal Highlands
21. East Coast Boreal Region
22. Boreal Lake Plateau
23. Whale River Region
24. Northern Labrador Region
25. Ungava Tundra Plateau
26. Northern Davis Region

HUDSON BAY LOWLANDS

27. Hudson-James Lowlands
28. Southampton Plain

ST. LAWRENCE LOWLANDS

29. (a) West St. Lawrence Lowland
    (b) Central St. Lawrence Lowland
    (c) East St. Lawrence Lowland

APPALACHIAN

30. Notre Dame · Megantic Mountains
31. Maritime Acadian Highlands
32. Maritime Plain
33. Atlantic Coast Mountains
34. Western Newfoundland Island Highlands
35. Eastern Newfoundland Island Atlantic Region

ARCTIC ISLANDS

36. Western Arctic Lowlands
37. Eastern Arctic Lowlands

HIGH ARCTIC ISLANDS

38. Western High Arctic Region
39. Eastern High Arctic Glacier Region

26

Auyuittuq

25

Torngat Mountains

23 24

22

21

Mealy Mountains

Gros Morne

29c

35

20

32

34

30

19c

La Mauricie

Cape Breton Highlands

33

29b

31

19b

29a

Fundy

Kilometers

200   0   200   400   600   800

thus apparent in three ways. First, the park was picked for its representative rather than its commercial or political value. Second, the CPS successfully insisted on prohibiting resource extraction within the park boundaries. Third, the park itself has been kept, at least so far, in a remarkably natural condition. Even after I had gotten directions from the modest visitor center, I drove past the trail head three times before hiking a mile up a dirt road to find it. I did find the one developed interpretive trail, but I did not see a single hard hat.

To say that the CPS is using preservationist ideals in attempting to expand the system should not be interpreted to mean that they are always successful. Two major obstacles have made the process often long and difficult. Both result from the preservation emphasis. First, as the Grasslands example shows, the local citizens are not always receptive to proposals for new federal parks or historic sites. This reluctance is intensified by CPS insistence on principles such as the banning of oil exploration or hunting inside the parks. Recently, for example, the proposal to establish a park near Churchill, Manitoba, has been questioned by local residents. The park, representing the Hudson-James Lowlands region, would contain polar bear denning sites. Many Churchill residents, particularly hunters and trappers, fear the effects of park protection on their pursuit of local wildlife such as bears. As one trapper said, "If it becomes a national park, it belongs only to the tourists."

The second obstacle to expansion is the cost of establishing parklands. Local reluctance and provincial negotiations slow the process, while the cost of lands and historic sites escalates. This problem is also a result of the emphasis on preservation of representative areas. The federal government, bound to its system plan, is forced to bargain with provincial governments for selected lands. Some observers feel that the pride that formerly stimulated provincial acceptance of national parks has been replaced by greed for federal dollars. Provincial leaders realize that they hold lands that have become precious because of their representative value. The latest addition to the national park system illustrates the point. South Moresby, representing terrestrial region one and Pacific Ocean regions one and two, was formally added to the system in 1988. The settlement followed long negotiations with the British Columbia government that were intensified by provincial concern for local logging interests. The federal government finally agreed to pay more than $25 million in compensation to loggers. The final cost to Ottawa of getting the area that the CPS recommended along with the prohibitions on logging is considerably more than $100 million, making South Moresby the most expensive Canadian park ever.

FUTURE.   The stated goal of the CPS is to "represent each of Canada's 39 terrestrial natural regions in the national parks system" and at least some of the marine regions.[57] The timetable for the future expansion of the Canadian park system is spelled out in the *Green Plan;* the government is to establish five new national parks and three new marine parks by 1996. By 2000 the government is to have completed negotiations on the thirteen remaining terrestrial areas needed to finish the system. In addition, fifteen of the historic themes listed in table 3-6 are to be represented by the end of the century. Overall, the goal is to "Set aside as protected space 12 percent of the country," with at least a quarter of that managed by the CPS as parks and historic sites.[58]

The delays in completing the park system have fostered skepticism among environmentalists and other observers. A mid-1980s task force headed by John Theberge of the University of Waterloo expressed public frustration with the slow pace of expansion and called for accelerated efforts in the next decade. Theberge recommended that some new parks might be run by the provinces or by native peoples. The criticism intensified when Mulroney's Conservatives were slow to add any new parks after being reelected in 1988. Environmental groups have delivered petitions and reports reminding the government of the *Green Plan*'s promises to finish the system. A joint effort by a variety of organizations under the title Canadian Environmental Advisory Council issued a comprehensive recommendation for system expansion in 1991. The report cites widespread public support for expansion: 60 percent of surveyed Canadians favored "at least doubling the amount of land protected as wilderness."[59] The report reiterates the plan for a complete terrestrial system by 2000 and a complete marine system by 2010. Finally, at the Fourth World Congress on National Parks in 1992, environmental groups led by Kevin McNamee of the Canadian Nature Federation publicly reminded Mulroney of his government's promises for park expansion.

The government claims to be on schedule. In 1992, Environment minister Charest admitted that obstacles were proving difficult to overcome but argued that the overall plans were still being pursued. The CPS, at least, is doing its part. In its *National Parks System Plan,* published in 1990, the agency describes each of the terrestrial regions and the status of proposed representative areas therein. Further, the agency calls for "a new consensus and determination on the part of all Canadians and all levels of government" to complete the system.[60]

In sum, comparison of the U.S. and Canadian procedures for system expansion are exemplified by Steamtown and Grasslands. Steamtown represents pork barrel politics; Grasslands represents a unique ecosystem. Simply comparing percentage of land protected is misleading. Roughly 3.6 percent of U.S. land is protected by the NPS. The Canadian plan calls for 3 percent of that country's land to be protected by the CPS when the system is finished. The Canadians, of course, have not yet achieved this goal. However, the method being used by the CPS to expand its system is dramatically different from that found in the United States. Comparative studies of world park systems have applauded the comprehensive and scientific approach now being used in Canada to preserve diverse natural ecosystems.

Erring on the Side of Preservation

Pukaskwa is a rugged park on the north shore of Lake Superior. Even in June, cold winds off the lake howl into the sixty-seven-site campground. The CPS supplies plenty of free firewood, but if that does not warm you enough, you will have to leave the park. Pukaskwa does not supply a hotel. Fifteen years ago, when the park was first established, plans called for a hotel complex. CPS managers decided, however, that a hotel would, as Superintendent Murphy says, "not be consistent with the park in a natural state." If developers come now with proposals, Murphy continued, the CPS position is backed by federal policy and official documentation. Murphy concluded his story: "From now on, we will always err on the side of preservation." Murphy's conclusion illustrates the recent reorientation of the agency's management policies to encourage use of the parks, but only in ways that do not disturb the ecological integrity.

PLANNING.   With the decentralized Canadian park system, knowing the person who writes a park's management plan provides revealing insight into the document. Paul Galbraith, Prince Albert's chief warden, is a man intensely concerned with environmental issues who is also currently overseeing the rewriting of the park's planning document. When I talked with Galbraith, our conversation ranged from specific park concerns to the writings of deep ecologist Stan Rowe. One of Rowe's statements gives a sense of the philosophical orientation: "Perhaps by saving wilderness we preserve an important part of ourselves. . . . The only firm foundation for wilderness preservation is psychological and attitudinal."[61] Galbraith's working version of the management plan reflects Rowe's influence, envisioning a future where park visitors "have improved their mental and spiritual health and general well-being. They apply what they have learned and experienced in the park to promote ecological harmony wherever they live. They understand more clearly who they are as Canadians and respect nature as a life-support system."[62] The intentions of Galbraith and other CPS personnel working on the document are not idle musings on spiritual well-being. These trail-level bureaucrats are designing an ecosystem approach for Prince Albert to allow visitation and at the same time to preserve the natural wilderness in the park.

Each Canadian unit uses a management plan, as do U.S. parks. The plans provide direction for park managers but also represent "commitments to the public . . . regarding the use and protection of national parks."[63] The plans, similar to the general management plans of the U.S. parks, identify types and means of resource protection and visitor service. The plans are prepared by each park's management team with input from the regional office and the public through workshops and reviews. Each park's plan is to be reviewed every five years.

The CPS also uses a system of zoning in parks. This system in itself reflects a change over the past fifteen years. In the previous policy statement, that of 1979, zoning was recommended as "a guide for the activities of both visitors and managers" but mandated to be used only "as appropriate."[64] According to observers, zoning was thus not formally instituted on a consistent basis. Today's policy uses planning zones for "direction for the activities of managers and visitors alike."[65] The zones are:

—Special Preservation: areas containing unique, threatened, or endangered features, where motorized access and sometimes even general access is prohibited;

—Wilderness: areas where only activities that "do not conflict with maintaining the wilderness itself" such as hiking are allowed, and even these may be limited;

—Natural Environment: areas in which low-density activities are allowed, and motorized transport for access is permitted but not encouraged;

—Outdoor Recreation: areas where a broad range of activities for visitor enjoyment and outdoor recreation is allowed;

—Park Services: Areas containing visitor services and support facilities including town sites.

The tone of the CPS planning policy statements is heavily oriented toward preservation. The second paragraph in the planning section cites the 1988 amendments to the National Parks Act requiring that "the maintenance of ecological integrity must be the first consideration in management planning." The second guideline for the planning process states: "The maintenance of ecological integrity through the protection of natural resources will be the first priority when considering zoning and visitor use."[66]

Prince Albert's revised plan is a good example in its emphasis on preservation and its heavy reliance on specific references to Canada's *Green Plan*. For example, the working document explicitly cites the *Green Plan* for its overriding philosophy: "We must think, plan, and act in terms of ecosystems."[67] The Prince Albert plan operationalizes this approach in the following ways. First, management decisions will be based on recognition that all parts of an ecosystem, including species (such as humans) and processes (such as fire) are interrelated. Second, actions will reflect an outward-looking perspective since "ecosystems do not stop at park boundaries." Third, the park will provide a forum for communicating to visitors the "knowledge, skills, and values" necessary for environmental citizenship. Fourth, park managers will continue to pursue and to encourage sound ecological research. Finally, the town site of Waskesiu Lake will be managed to preserve the cultural heritage and to display an environmental lifestyle.

Several comparative dimensions of planning are apparent. First, the tone of the NPS planning guidelines, as in the cited statement on the National Environmental Policy Act, is noncommittal in establishing priorities between use and preservation. The planning section in the CPS guidelines is decidedly preservationist in tone and substance. Second, the CPS section shows more agency autonomy. While the NPS includes congressional directives in its planning section, parliamentary involvement in CPS plans is only mentioned in legislated emphasis on preservation policies and in the approval of already formulated plans. Third, the zoning systems are comparable but not identical. The CPS uses a zone where access may explicitly be prohibited, whereas the NPS does not. Further, the CPS has no provisions for activities such as those listed in the special use zone of the NPS. Fourth, while the NPS statements ignore the Wilderness Act, the CPS guidelines emphasize the use of similarly designated wilderness areas, particularly in the categories of zones one and two. Finally, the relative autonomy of CPS employees such as Galbraith in writing or rewriting park planning documents is readily apparent.

VISITOR USE.   Point Pelee is a bird-watcher's paradise located on the southernmost point of the Canadian mainland. Situated along migratory routes, the park's 3,500 acres stretch out into Lake Erie, providing a last rest stop for birds flying south and a landing strip for those coming back north. Over 300 species have been officially counted in the park. Until recently, one winged creature was subject to more than just watching. Beginning in 1885, the park hosted an annual duck hunt. On June 6, 1989, the CPS officially ended the hunt, the last remnant of sport hunting in Canadian national parks.

The experience at Point Pelee illustrates CPS policies toward visitor use. The agency is well aware of the need to sustain levels of visitation and thus does not discourage tourists. However, the CPS has become more preservation-oriented in recent years in controlling what those visitors do while they are in the parks.

In discussing the recent shift of the CPS toward preservation, I do not mean to suggest that the agency attempts to lock up natural areas. The CPS explicitly welcomes visitors to the parks. Policy guidelines state: "To keep faith with their owners, the people of Canada, the CPS must ensure that such opportunities remain forever available."[68] The agency has developed fairly sophisticated models to determine socioeconomic impacts of parks and to facilitate the incorporation of public concerns into park management.[69] For example, the Visitor Activity Management Process (VAMP) has been used to assess visitor demands on specific parks and to identify the appropriate mix of activities within each unit. As a result the agency produces sophisticated documentation to show how park units contribute to the economic well-being of the tourist industry in general and to the economies of the specific provinces in which park units are located.

To illustrate, consider the effects of visitor use of the national parks in Alberta. These parks provide an important illustration because they are the most popular and best known, but also because they are located in a province where powerful economic interests often advocate other uses, such as oil development, for land. In fiscal year 1987–88, Banff, Jasper, Elk Island, Waterton Lakes, and Rocky Mountain hosted more than 6 million visitors. Roughly 80 percent of these visitors were Canadian, but more than 40 percent of them came from outside the province. The CPS thus used these figures to show Albertan authorities how much money comes into the province from outside. Visitor expenditures associated with the parks totaled more than $412 million. In addition, the Parks Service spent more than $54 million to maintain the parks. These figures represent about 14 percent of the gross of Alberta's third largest industry, tourism. As the CPS says, "The operation of parks and the delivery of the program managed by the CPS in the province of Alberta results in significant benefits accruing to the economy of that province."[70]

The CPS continues to encourage high levels of visitation and uses figures such as those in Alberta to tout the economic justifications of parks. Like the NPS, therefore, the CPS has not yet limited visitation by using a quota system for entrance into any national park. Further, the CPS has not yet installed a comprehensive reservation system for camping like that in the United States.

Given the emphasis on maintaining visitation levels, however, the CPS has concentrated in recent years on controlling how visitors use the parks. The shifting emphasis toward preservation is evident in many ways. While the CPS has not adopted an entrance quota system, it did, in the 1980s, begin establishing trail and campsite quotas and some specific reservation systems. In some places, automobile traffic is banned. At Point Pelee, for instance, visitors wishing to see the point itself must either hike, bike, or ride the nonpolluting trams out to the trail head. Official policy encourages combining environmental education along with visitor use. In fact, the policy section on visitor service is entitled "Public Understanding and Enjoyment of National Parks."

Recent changes are particularly evident with modifications to traditional forms of recreation. Hunting in national parks has recently been curtailed. Termination of the duck hunt in Point Pelee illustrates that the CPS is willing to sacrifice potential revenue in order to make changes. Now CPS official policy 3.1.4 states flatly: "Sport hunting will not be permitted in a national park." Park wardens also instituted various programs to crack down on illegal hunting in the 1980s. The battle against poaching now involves a systematic CPS effort using a computerized information center and the latest crime-fighting techniques.

The changes in management of sport fishing illustrate the increased emphasis on preservation and the effect of input from the field. In 1989 Prince Albert warden Galbraith wrote a piece on park aquatic resources in which he recommended that procedures be changed to modify angling to make it "part of an overall national park experience rather than a traditional use."[71] Two years later the CPS policy guidelines echoed Galbraith's suggestions by imposing regulations that "conform to the principle that angling is part of an overall aquatic resource program involving public education, recreation and resource protection."[72] These changes include discontinuing fish stocking except where needed to restore indigenous fish populations and encouraging increased use of catch-and-release fishing.

Other policy modifications of recreational activities began at the field level and then expanded into systemwide practices. After numerous battles in the field against expanding ski resorts, 1991 CPS policy prohibited development of new ski areas or expansion of existing areas. After Banff managers indicated no interest in managing the park's golf course, 1991 CPS policy barred any new courses in national parks. Field managers have also cracked down on the use of off-road vehicles. Management plans in the mid-1980s for the mountain parks banned recreational snowmobiling. Official CPS policy now states: "Non-motorized means of transportation will be used in national parks wherever feasible."[73]

Though the CPS has changed many of its policies on recreation in recent years, one policy area has remained the same. Indigenous people are still allowed to continue their traditional practices of hunting, fishing, and trapping in the new northern parks. This policy will continue in the future. A recent proposal seeks to amend the National Park Act to allow traditional trapping, hunting, and fishing in Wood Buffalo.

In comparative terms, contrasts between the CPS and the NPS in agency control of visitors are more apparent than in encouragement of visits in general. Neither service has yet imposed quotas on entrance to national parks. Once visitors are in the parks, the NPS continues to allow use of resources through diverse means. The CPS has moved toward preservation of resources by restricting or even banning many of these activities. The CPS changes have obviously been driven by field employees in recent years.

COMMERCIAL USE. The newly elected president of the Jasper town site Chamber of Commerce gave me a different perspective from that of CPS employees on the commercial use of national parks. Citing the historical experiences of three generations of hotel owners, George Andrew criticized the parks agencies for sometimes being too inflexible in dealing with town sites. Andrew added, however, that the relationship has been more cooperative in the past few years. The reason, he said, was not that the CPS managers had been coopted, but rather that citizens in Jasper have realized that "most sensible people would say there is a limit to what an area can handle." Agency management of town sites like Jasper and of other commercial activities affecting the parks is still evolving, but in directions that reflect concern for natural resources and conditions.

While the parks agencies encouraged the growth of town sites for decades, they began to change their management of them in 1979. The 1979 policy proposal included, for example, a "need to reside" provision, which determined that residency in town sites would be "limited to those who are providing essential services and who cannot reasonably live outside the park."[74] Controversies about the town sites continued into the 1980s. Townspeople and park officials often clashed over maintenance of facilities, development plans, airstrips, and cost-cutting proposals such as closing highways in winter. The resentment of CPS employees toward unnatural settlements within parks has fostered some acrimony and some new policies. Today the CPS states emphatically: "No new communities will be developed within national parks."[75]

Policies for existing town sites are more ambiguous. In its latest policy statements, the CPS recognizes and classifies the town sites at Banff and Jasper as actual towns, communities qualified to support local self-government. Town sites in Prince Albert (Waskesiu), Riding Mountain (Wasagaming), Waterton Lakes (Waterton Park), and Yoho (Field) are still to be managed by the CPS as a "focus for and concentration of visitor activity services and facilities."[76]

The tensions between townspeople and park managers have moderated somewhat recently. In part this moderation is due to the realization of residents like Andrew that unlimited growth could ruin the pristine qualities that draw tourist business and thus result in less satisfactory living conditions. Trail-level bureaucrats have also displayed a moderation in attitudes. Park officials are now more likely to work with residents to control effects than against townspeople by moving them out. The prototype for this approach was the management planning process for the mountain parks that used citizen input in choosing among several different options for future development of town sites. The plan calls for improving existing facilities rather than building new ones, repaving existing roads but not laying new ones, and filling in town sites like Banff rather than allowing urban sprawl. As Environment minister McMillan said of the use-preservation mix in the plan, "When there is a conflict, however, between the two, the bias is distinctly in favor of preservation values."[77]

In addition, CPS employees have become resigned to the existence of the town sites. While managers will readily admit that towns are not what you expect to find in the parks, they now put a good face on their presence. Prince Albert superintendent Dalman, for example, refers to his park's Waskesiu town site as a "cultural artifact that will never be removed and, because it is an artifact, perhaps shouldn't be." Instead of trying to remove it, Dalman and the CPS are now attempting, as the planning documents make explicit, to make the park's town site a "model environmental community."[78] To cite another example, Yoho's managers are building a learning center in Field to provide modern environmental education to visiting students in ecologically designed buildings that are three times as energy efficient as the structures they are replacing.

The CPS, notorious in previous decades for encouraging other commercial uses of the parks, changed many of its policies in the 1980s. The policy on extraction in new parks is quite clear: "Commercial exploration, extraction or development of natural resources will be terminated before national parks are formally established."[79] This 1991 policy marks a final repudiation of the statements of Minister Blais-Grenier, who in 1985 promised new parks with compromises such as provisions for logging and mining. CPS policy on air travel in new and established parks is also explicit. "Access by private or commercial aircraft within national parks will not be allowed except to remote areas where reasonable travel alternatives are not available."[80]

The policy on commercial extraction from existing parks is more complicated. Certainly, the logging situation in Wood Buffalo has attracted considerable attention. Now that it has been resolved, the agency has promulgated a tougher line on such uses in existing parks, making it clear that only the "traditional subsistence uses" by indigenous people may legally continue in parks.[81] One could argue that talk is cheap, and that the CPS will not easily abandon old commercial practices. In the area of logging, however, revenue figures again reveal the agency's commitment to change. Agency revenue from timber permits was cut nearly in half between fiscal year 1989—90 and 1990—91. Of the $105,204 remaining in timber permit revenue, nearly all ($104,606) came from Wood Buffalo. All other park units and regions reflect a steady decline in timber operations.

Cows continue to graze on NPS land, but CPS policy now prohibits them from new parks and demands their removal in most cases from existing ones. NPS policy still explicitly allows commercial uses of park space such as grazing, mining, logging, and air travel. CPS policy has shifted away from this use and toward preservation of the resources for their own value. Much of this change in CPS behavior was driven from within.

FIRE MANAGEMENT.   The 1988 Yellowstone fires altered CPS policy just as they changed NPS policy. The policies changed in different directions. The Canadian response to the Yellowstone inferno was perhaps best summed up by the Banff superintendent, who said parks like Yellowstone had created their own risks through 100 years of mismanagement by not allowing natural fires and not using planned burning.[82] Yellowstone managers like Robert Barbee also knew this, but their arguments were overruled in subsequent policy discussions. In Canada input from field personnel had a significant role in changing fire-management policies.

The changing CPS fire policies show a dramatic evolution. Before 1979 Parks Canada practiced suppression of all fires. Following the U.S. shift to allowing natural fires in 1972, Canadian fire policy was criticized for its continued emphasis on limiting fire. The policy was changed in 1979. "Manipulation of naturally occurring processes such as fire . . . may take place only after monitoring has shown that" fire will involve such outcomes as adverse effects on other lands, threats to public safety, and threats to park facilities.[83] The agency then instituted procedures similar to those in Yellowstone between 1972 and 1987 involving occasional letburns and limited use of heavy firefighting equipment.

After the Yellowstone fires, the CPS recognized and admitted its own past mistakes in not using natural fires and prescribed burning properly as a prelude to instituting a "comprehensive fire management program."[84] When the CPS issued its new policy guidelines for comment in 1991, the changes in fire policy were apparent. The new section 3.2.4 states: "Manipulation of naturally occurring processes such as fire . . . may take place for reasons other than the achievement of park resource management objectives when no reasonable alternative exists and it is clear that, without intervention, there will be serious adverse effects on neighbouring lands or risk to major park facilities, public health or safety."[85] The phrase "after monitoring" in the earlier policy has been replaced by "when no reasonable alternative exists" in the new policy. The change in emphasis is more than just semantic. The current program is built around using planned and naturally occurring fires to "help attain better ecosystem balances" and describes the deliberate use of prescribed fire as "a critical element."[86]

In a comparative sense, the NPS must assume risk to people and property and then show that the fire will not have adverse effects, while the CPS is to assume that people and property will not be hurt and will allow fires to burn until it is proven that harm will come. The NPS policy, determined largely by USDA officials and senators like Alan Simpson and Malcolm Wallop, emphasizes safeguards on persons and property within parks. The less-restrictive changes for the CPS responded to the perceptions of scientists and managers in the field, who argued that wilderness areas could best be preserved by allowing natural processes such as fire.

SCIENCE AND RESEARCH.   During our discussion Banff superintendent Zinkan was describing some of the external threats to the park. He looked at the map in the management plan that showed the interior of the park in some detail; beyond the park boundaries was blank space. "What nerve we have," he sighed, "to draw the map as if there's nothing else around us." Zinkan's disgust reflects a new attitude among CPS personnel, an attitude that necessitates directing science and research programs at surrounding areas as well as those within park boundaries. CPS employees intend to use research efforts within an ecosystem approach that will no longer consider parks as environmental oases surrounded by deserts of resource development and use.

The early history of Canadian attention to science and research in the parks parallels that of the United States. Before the 1980s, Canadian parks agencies were criticized for the lack of ecological research and planning despite identified and well-recognized needs. Ecological principles were often misunderstood or avoided. As one observer concluded in 1970, "Present research efforts are very small in comparison with the importance of the problem."[87] Even as late as 1985, the Assembly Conference recommended greater attention to research efforts.

Changing emphasis on science in recent years has been reflected in the agency's policies. By 1979 the CPS was calling for resource databases in each park, reflecting an awareness of the need for information but retaining the decentralized nature of information collection. The 1979 policies also stated: "Where active resource management is necessary, techniques will duplicate natural processes as closely as possible."[88] The 1991 version of that same paragraph reflected increasing attention to the potential use of scientific tools: "Where active resource management is necessary and practical, it will be based on scientific research, use techniques which duplicate natural processes as closely as possible, and be carefully monitored." The government's Green Plan also endorsed the use of "environmental science" to "explore new directions in solving our domestic and global problems." As with the other policies, increased awareness of potential use of science was present throughout the organization. The Western

Region's Task Force on Science and Protection concluded in the year before the new CPS policy: "The challenge is to increase the amount of decision-making inputs that are based on scientific natural resource and socio-economic data and analysis while reducing those dependent upon intuition."[89]

Like the NPS policies on science, these statements could be only words reflecting few specifics and little commitment. The CPS still contracts most specialized scientific efforts out to other agencies, universities, and consultants. On the other hand, the agency speaks with pride about specific projects such as the previously mentioned socioeconomic analyses and the increased use of electronic data processing like geographic information systems. Further, while the NPS explicitly focuses on in-park issues, the CPS endorses a broader view of its scientific efforts. The 1991 policy guidelines state: "Scientific research other than that which is essential for park management but which contributes to the protection and public understanding of ecosystems will be encouraged, including the use of parks as benchmarks for ecological research."[90] One example is the continuing effort in La Mauricie to study the effects of acid rain by monitoring the acidification of lakes within the park.

Official endorsement of greater attention to science and research is a recent phenomenon in both agencies. Both currently espouse the use of more data and more sophisticated analyses in making decisions. The most noticeable difference between the two agencies' statements is explicit CPS awareness, compared with NPS focus on in-park efforts, that the use of science for preservation requires "understanding and management of ecosystems both inside and outside national parks."[91] The willingness to address external issues might make a difference in future situations.

EDUCATION.   Imagine the following situation. A park has an elk population that is constantly in potential danger from automobile traffic. Collisions, which are frequent along one particular stretch of road, result in vehicle damage and elk fatalities. How would you address this issue? When Prince Albert chief warden Galbraith asked me this question, he also read my mind. Most people would say—he answered his own question—Put a fence up to keep the animals from crossing the road. "I say," he went on, "educate people to drive slower." One is a technological fix that separates people from the ecosystem while still allowing them to use park resources as they wish. The other minimizes human impact in the long run to preserve the natural setting. In Canada trail-level bureaucrats such as Galbraith intend to use environmental education both for and with the parks.

The CPS mandate explicitly includes education as one of the purposes of park units. Early educational efforts by the parks were directed inward, almost exclusively to informing visitors about what they would find within individual units. In 1979 the agency officially recognized its "responsibility to inform the Canadian public about their national parks and to provide programs which encourage a better understanding of these natural areas of Canadian significance," suggesting an expansion of educational programs.[92] Nevertheless, the Banff Centennial Assembly Conference demanded in 1985 that environmental education be made a priority.[93] A task force as late as 1986 found the agency having difficulty shedding its own fortress mentality to adopt an educational leadership role in the community.

In environmental education efforts, a shift is apparent in official statements of policy. As in other policies, the shift comes from all levels of the organization. The 1990 Strategic Plan calls for employees to "Promote and advocate the philosophy of environmental stewardship" by increasing public awareness of the benefits of environmental protection through high quality education programs directed at a broad audience.[94] Both the *Green Plan* at the federal level and the Parks and People Task Force at the regional level call for replacing the fortress mentality—which in the past prevented using education on issues arising outside park boundaries—with a focus on using the parks to exemplify environmental stewardship. The 1991 CPS policies call for extending education programs outside the parks to include "cooperative action with other agencies, groups and individuals across a broad spectrum of interests."[95] The CPS has officially endorsed using expanded educational efforts to preserve parks from external as well as internal threats. At the level of the parks, trail-level bureaucrats are putting theory into practice. The Prince Albert management plan, for example, calls for using the park to develop "an environmentally literate society."[96]

As in the United States, historical education is facilitated through the preservation and presentation of historic sites. These sites are designated for inclusion in the system by the Historic Sites and Monuments Board. Particularly significant sites are designated as national historic parks. These sites illustrate important historic periods or events through reproduction in a form as authentic as possible of living conditions, costumes, and structures of the represented period. Official policy calls for the CPS to preserve or restore these sites to "authentic historic form with a minimum of modification to suit human needs."[97] This mandate precludes towers such as that found at Gettysburg. The Canadian government has devoted substantial effort and resources to these sites in the past two decades.

The two park service agencies have increased attention to education in recent years. As with science policy, however, the increased emphasis also involves shifts in focus. The NPS, by its own admission, continues to educate inward, while the CPS, at least in policy statements, has expanded its educational efforts outward.

Taking It and Running

In Canada, empowerment of trail-level bureaucrats and an increasing focus on preservation go hand in hand. When I asked Western Region official Jillian Roulet if these two trends were consistent, she answered: "They weren't necessarily planned that way, but they do reinforce each other." More than one CPS field manager echoed her next comment: "We're going to take it [power] and run with it." Her comments suggest the potential impact of the human resources within the CPS.

PERSONNEL.   The CPS employs 4,700 full-time-equivalent workers. Most CPS employees are on a government pay scale, but the average salaries are somewhat higher than those in the NPS. For example, the average annual pay for a field warden in 1991 was $41,357. Even taking into account the lower value of the Canadian dollar (U.S. $1.00 roughly equals Canadian $1.20), the CPS pay scale represents substantially higher compensation than that of NPS employees.

In part because of adequate compensation, competition for CPS jobs is still fierce. Since 1986, qualifications for new hires have included a high school degree and at least two years of postsecondary training in natural resource management or one of the natural sciences. Turnover of personnel is low, less than 5 percent in most regions. Ironically, the CPS is facing potential problems with large numbers of eligible retirees in coming years precisely because it has been so good at retaining employees. In the Prairie Region, for example, 20 percent of the work force will be eligible to retire in the next decade. Morale, except for the fear of potential layoffs described earlier, remains high.

Training and career paths for CPS employees have traditionally been as nonsystematic as those of their U.S. counterparts. Careerists can get courses at the Royal Military College in Kingston or at field offices in both eastern and western locations. Like the United States, however, attendance is somewhat random. In addition, course emphasis has been on leadership and protection rather than on resource management. Little uniformity in training exists. Human resource planning in the organization has histori-cally been "about zero," as Prince Albert superintendent Dalman told me. Career paths have been roughly comparable to those in the NPS. Most managers rose up through the ranks, and most have backgrounds in visitor service rather than in resource management.

Nevertheless, some differences exist and some are now occurring that preview a shift consistent with the hypothetical turn toward preservation. First, unlike the United States, regional staff officers in the CPS are not required to have previous management experience at the field level. At first glance this may appear to create potential for moving away from field

TABLE 3-7. *Budget Allocations of the CPS, 1991*
Millions of Canadian dollars unless otherwise specified

Item	Percent	Allocation	Appropriation
Operations/maintenance	19	77	408
Visitor service	16	65	408
Resource management	9	37	408
Park supervision	6	26	408
Research	6	23	408

Source: Environment Canada (1992, pp. 4-10, 4-27, 4-40).

concerns. In fact, it enables the recruitment of people with fresh perspectives in staff positions. This capability to hire regional officers from the outside enabled the Western Region to bring in the dynamic Sandra Davis, for example. In addition, regional offices can increase concentrations on certain perspectives. Recently the Prairie and Western regions have hired more specialists in resource management. Hiring of outsiders in these positions has not fostered the resentment seen in the NPS from field employees toward noncareerists. Outsiders in the NPS have centralized power, while those in the CPS, like Davis, have encouraged reliance on field input.

Second, the CPS has recently begun emphasizing the need for more training in resource management and other relevant sciences. As mentioned, new hires are expected to have some training in natural sciences. In addition, *Public Service 2000* calls for greater commitment of funds and training to develop expertise in new areas, such as resource management, for agency personnel. These recommendations have been endorsed by interviewees at all levels of the organization and in agency documents.

Third, the Western Region has already implemented changes, as shown in figures 3-3 and 3-5, reflecting a greater emphasis on preservation. The Warden Service is now termed the Department of Heritage Resource Conservation. Regional officers say the change is more than just a difference in title; it is intended to reflect more emphasis on science and resource management. The region has also combined interpretation and visitor services into heritage communication so that information conveyed to visitors is more than just responses to visitor demands but also planned environmental communication.

Neither the NPS nor the CPS has shown strong tendencies in the past toward fostering preservation-oriented skills in its human resources. In recent years, however, a divergence has become apparent. The NPS continues to keep compensation low, to avoid educational requirements, and to reject attempts to change career paths or agency emphases. The CPS has raised compensation, imposed educational requirements, and endorsed the personnel changes that are occurring at lower levels of the organization.

TABLE 3-8. *OLS Estimation of Park Unit Appropriations*

Independent variable	Coefficients
Constant	2.52
	(.03)
Visitors	0.73[a]
	(3.38)
Years	−7.57
	(.45)
Size	1.38
	(.39)
Threat	3.45[a]
	(2.71)
N (number of cases)	31
R²	.52

Sources: Dependent variable (FY89 expenditures), Canadian Parks Service (1989c); visitors (visits per year), years (years before 1990 park estimate), size (total square kilometers), and threats (to park units), Canadian Parks Service (1990e).

a. Significant at the .05 level.

BUDGETS. Table 3-7 shows how the CPS has allocated its funds among its various functions in recent years. While this table invites comparisons with its counterpart in chapter 2, differences in the accounting and reporting systems of the CPS and the NPS make it necessary to use care in evaluating these figures. The table shows the latest available data, for fiscal year 1990–91, for the CPS. The numbers in the table reflect percentages of total expenditures by the agency. The numbers in column two are the actual figures. As with the NPS data, I calculated the percentages in adjacent years to ensure that these figures were representative. The percentages were quite similar. The percentages do not total 100 because of all the other categories such as land acquisitions that are not included.

Two aspects of table 3-7 stand out. For the most part, CPS and NPS expenditures are extremely comparable. Both agencies spend about 10 percent of their budgets on resource management, 16 percent on visitor service, and 6 percent on park supervision. The 16 percent figure is somewhat misleading because it represents only direct expenditures on visitor activities and not indirect costs. The operations maintenance figure of 19 percent is roughly comparable to the 23 percent maintenance figure for the NPS. The research differential is consistent with the hypothetical swing of the CPS toward preservation. The CPS currently spends roughly twice as much of its budget on scientific planning and research as does the NPS.

Table 3-8 presents an analysis of allocations to individual park units. The analysis uses the same econometric method applied to the NPS in chapter 2. As with the NPS, the allocation equation is quite strong, with multiple significant independent variables. The two variables that are significantly related to the allocation of funds to park units for both systems are visitors and threats. These relationships suggest that unit allocations are determined by how many visitors a park entertains per year *and* the number of threats it must manage. Little differentiation is apparent between the two systems in unit allocation.

Some caveats about the CPS equation are in order. The data for the CPS only represent thirty-one cases. CPS employees are currently developing more sophisticated means for counting visitors and other variables. Data are therefore only available in limited amounts. As a result, the CPS equation is less reliable than that for the NPS. Further, the data are not as clean as desirable. For example, the visitation figures and the appropriation figures for the CPS are from the same year. To be more accurate, one would need visitation data from years preceding the appropriation to allow for the lag time in fiscal determinations. This lag time is used in the NPS equation. In other words, one can only take these results as illustrative of the point, not as definitive analysis.

Certain reasons for the lack of differentiation between the two systems are important. Again, timing is crucial. The shifts in both agencies have occurred only recently, whereas spending decisions are the culmination of years of tradition. This is especially true in the CPS, where changes are continuing and recent. Thus insufficient time has passed to show dramatic changes. A second major reason for inconclusive quantitative evidence is the multicollinearity in the equation described earlier. In both systems, threats are related to number of visitors.

The quantitative evidence on budgets is not conclusive regarding the hypothesized shift of the NPS away from and of the CPS toward preservation. Nevertheless, certain aspects of program and unit allocations support the hypothesis. The CPS spends more on research than does the NPS. The NPS spends substantially more on maintenance for visitors than does the CPS. More important than aggregate amounts is how the appropriated money is spent. To understand these decisions requires examination of programs in the field, the subject of the next chapter.

Summary

The CPS is still evolving into a different organization than it was fifteen years ago. Changes in the political environment have enabled CPS employees to decentralize their agency and shift the focus of policy behavior. As Western Region official Roulet said proudly, "Used to be, people would talk about the dual mandate. Now they say preservation first and then use if the parks can handle it."

Chapter 4

Being Smart

"There are a lot of things we can do in the parks to show how we can be smart."

— *Yoho Superintendent Ian Church*

Superintendent Church had definite views of what both "we" and being "smart" meant. He had just described an incident at Lake O'Hara, the most acclaimed destination in Yoho National Park. Visitors need reservations months in advance to camp or to stay at the lodge there. The Canadian Park Service (CPS) already imposes limits on visitation and use, but on the preceding weekend, with beautiful weather and a full load of visitors anxious to go hiking, Church had shut the area down. A grizzly bear, a threatened species, had wandered into a campsite looking for food. To protect the bear, the CPS required people to stay inside the lodge from which they were bussed out to designated hiking areas, thereby altering the weekend plans of dozens of visitors. Even more remarkable is that, according to Church, the visitors were "one hundred percent supportive." "It's an indication of the future," Church said, "we will protect animals." Church is referring to the field employees of the CPS. Being smart means allowing only use of parks that does not compromise the preservation of natural features such as grizzly bears.

Analyzing changing agency behavior at the field level is important. Moving decisions and responsibilities down to the field level does not preclude political influence. Organization theorists have pointed out that decentralization does not translate to autonomy from politics. Indeed, as some have argued, "A decentralized organization with operators working alone in isolated outposts might well have decided that its task was to please whatever dominant and politically influential group existed in local communities."[1] Kaufman's seminal account of the Forest Service described the efforts of the agency, such as routine transfers, precisely to avoid this "capture" by local interests.[2] With the park services, local capture might translate to autonomous managers virtually turning over parklands to business, commercial, and local development interests. Central decisions, on the other hand, could be made by powerful, preservation-oriented, environmental interest groups.

My analysis suggests quite a different outcome. I argue that CPS trail-level bureaucrats like Church have used their recent empowerment to push the agency toward emphasizing the preservation of natural conditions. NPS field employees are likely to have their policy preferences overruled by political authorities whose electoral and ideological concerns often prevail over concern for natural resources.

Wildlife Management

In recent years, the National Park Service (NPS) has been forced to give visitor service a higher priority than wildlife protection, while the CPS has moved toward preserving endangered species in natural ecosystems. The differences are apparent in Yellowstone and Elk Island.

Why You Will Not See Grizzly Bears in Yellowstone

I was in Denali when I first saw the world's most awe-inspiring animal in a wild setting. Three grizzly bears were calmly eating berries about one hundred yards from where I stood. Having hitchhiked from Indiana to Alaska to see these creatures, I was, needless to say, moved. I have been fortunate enough to see grizzlies since in Jasper and Waterton Lakes. I have never seen them in Yellowstone. And, even though I plan to return many times, I doubt that I ever will.

People go to Yellowstone for many reasons. The world's oldest park is rich in history. It was first seen by a white man when John Colter left the returning Lewis and Clark expedition just days short of civilization so he could return to the mountains. Yellowstone's beauty is abundant. Even the tremendous fires of 1988 left dramatic vistas of burned tree spars and scarred hillsides. The park is also famous for its wildlife, especially its bears. How many of us who traveled through the park years ago have home movies or photographs of bears ambling up to our cars or into our campsites to look for food?

Today you might see a black bear if you are lucky. Your chances of seeing a grizzly are about as good as of winning the lottery. Grizzlies are the larger and more ferocious cousins of black bears. In many ways they are among the most magnificent animals still roaming the earth.

The wildlife policies of recent years have made sightings of grizzlies rare occurrences. The policies result from decades of bear mismanagement in Yellowstone and from recent political decisions. The NPS made mistakes on its own, but, for the most part, they were mistakes of ignorance. Recent political calculations by central authorities, in spite of advice and NPS

TABLE 4-1. *Grizzly Bear Statistics in Yellowstone, 1973–90*

Year	Females with cubs	Grizzly bear mortality	
		In ecosystem	*In park*
1973	14	. . .	. . .
1974	15	. . .	. . .
1975	4	. . .	. . .
1976	16	. . .	. . .
1977	13	11	3
1978	9	9	2
1979	13	10	1
1980	12	6	2
1981	13	6	0
1982	11	13	3
1983	13	6	1
1984	17	8	3
1985	9	6	1
1986	25	9	2
1987	13	. . .	. . .
1988	19	. . .	. . .
1989	16	. . .	. . .
1990	24	. . .	. . .

Sources: U.S. National Park Service (1988a, pp. 162, 163); Interagency Grizzly Bear Study Team (1991, p. 6).

intentions, have contributed to the demise of the bears. Those policies are apparent in the two park settlements of Grant Village and Fishing Bridge.

BACKGROUND.   Under the heading "Where Are the Bears?" the Summer, 1992, issue of the park newspaper reads, "In the past, bears were a common sight in Yellowstone National Park." In fact, early NPS policies made sure that this was so. As early as the 1890s, black and grizzly bears were common sights at park garbage dumps. The NPS even set up bleachers around these areas where tourists could watch the animals feed. Early in this century the number of grizzlies in the United States alone may have reached 100,000, and many of them were in Yellowstone. From censuses taken at the dumps and elsewhere in the park, scientists estimated that by the late 1960s, the park still held hundreds of grizzlies and even more black bears. No one knows for sure how many grizzlies are left today, but, as table 4-1 shows, the crucial number of cub-bearing females is estimated at two dozen or fewer.

The grizzly populations have been reduced because of diminished habitat, increased human interactions, and agency policies. In 1968 the

NPS began intensive garbage management, closing the dumps entirely by 1970 so that the bears would return to natural feeding. While the change in policy was intended to help the grizzlies in the long run, some scientists, such as the Craighead brothers, warned that abrupt changes would cause problems for the bears. The Craigheads, basing their recommendations on twelve years' study of more than five hundred grizzlies, urged supplementary carrion feeding in remote areas to assist the animals' transition back to natural feeding patterns. The NPS refused, and, without such feeding, the grizzlies began to roam in their search for food. Their wandering led to increasing human-bear interactions, which often resulted in troublesome bears' being destroyed. Many bears roamed outside the protection of park boundaries and were killed accidentally or hunted legally. A 1974 National Academy of Sciences study reported that 189 Yellowstone grizzly bears were confirmed kills between 1968 and 1973. Even the NPS admits that its own control procedures killed forty-eight grizzlies between 1968 and 1972.

During the 1970s, many associated with the NPS realized that bear management required more than just the "natural-feeding" policy. Future policy directions were unclear, however. As late as 1977, the twenty-two NPS areas with bear populations spent eight times as many days on garbage-handling activities for bears as on research for bears. The high mortality rates in Yellowstone in the years after 1968 led scientists and environmental groups to argue that grizzlies, in particular, needed legal protection. The Fish and Wildlife Service classified grizzlies as a "threatened" species in 1975, a status somewhat less protective than "endangered" but still demanding federal attention. Bear policies had already begun to change, however. Yellowstone's changes centered on bear-human interactions at the Grant Village and Fishing Bridge areas.

RECENT ACTIONS.   As figure 4-1 shows, if you drive into Yellowstone from the south, you pass through the Grant Village area. If you drive in from the east, your path takes you through Fishing Bridge. Both areas have provided visitor services: gas stations, stores, visitor centers, and campgrounds. Both settlements are geographically strategic, located at the intersections of major entrance highways to the park's Grand Loop Road. Both areas were at one time prime grizzly bear habitat. The bears were attracted to the areas because of human use and associated food sources. But both areas also contain natural features that make them valuable to bears. NPS biologists have described Fishing Bridge as important "because of the merging of lake, river, and terrestrial ecosystems that creates a diverse natural complex."[3] Grant Village was built where five trout streams enter Yellowstone Lake, making it also a desirable habitat for grizzlies.

FIGURE 4-1. *Boundary Map, Yellowstone National Park*

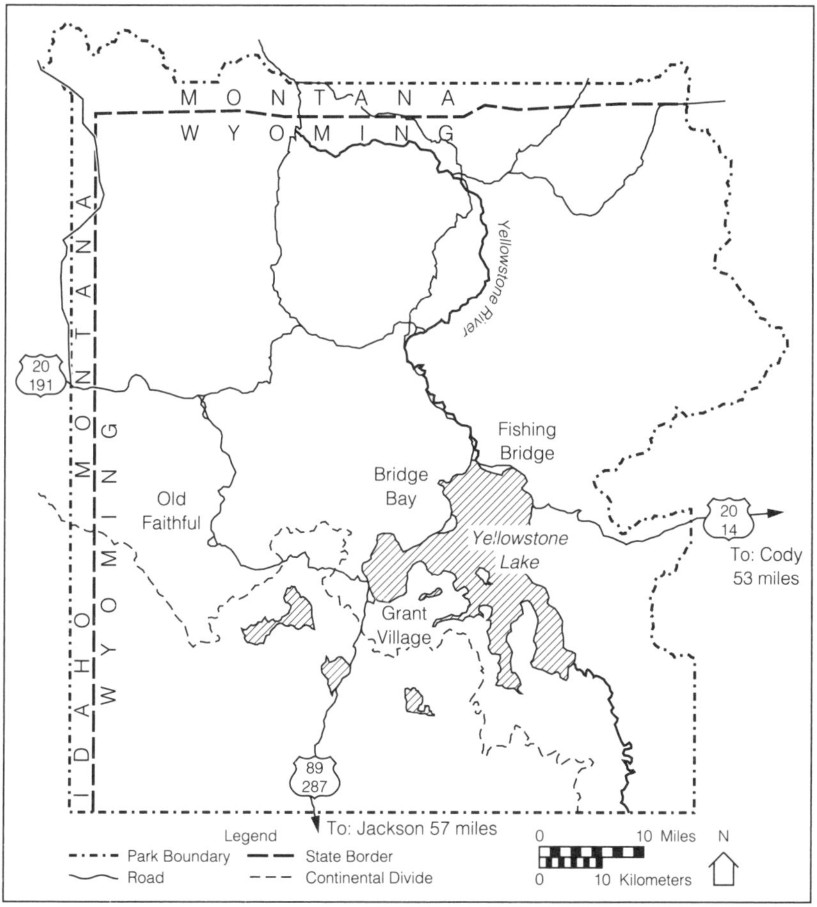

Source: U.S. National Park Service (1991d, p. 3).

Nevertheless, both settlements exist today as the result of a questionable compromise that was made worse by political intervention. By the time the NPS issued its master plan for Yellowstone in 1974, Fishing Bridge had been the site of human activity and thus of human–bear interaction for decades. Grant Village, meanwhile, was a planned development that had already been scuttled several times. In its *Master Plan,* the NPS traded one for the other. "Fishing Bridge: Current planning proposes to ultimately relieve congestion and eliminate accommodations and services from this existing developed area in order to facilitate restoration of critical wildlife habitats at Yellowstone Lake's outlet." Many of the existing Fishing Bridge facilities were to be moved to nearby developments, including Grant Village. "Grant Village," the *Master Plan* asserted, "will become a major development, containing several classes of accommodations."[4] Only the latter half of the bargain has been kept.

The next stage in the process revealed that decisions were out of park managers' hands. Since the grizzlies received threatened status in 1975, the plans for the two settlements required consultation between the NPS and the Fish and Wildlife Service. In 1979, the Fish and Wildlife Service issued the opinion that "we would not expect the new development at Grant Village to jeopardize the species if current management continues and facilities at Fishing Bridge are eliminated."[5] The Fish and Wildlife Service thus urged the NPS to honor its own compromise. Yellowstone superintendent John Townsley responded to the Fish and Wildlife Service, "At the park level we attempted to close Fishing Bridge Campground. National Park Service policy consideration *beyond Yellowstone* [emphasis added] resulted in the campground remaining open."[6] The NPS still planned to close the 300-site campground and the 350-unit recreational vehicle (RV) park at Fishing Bridge by 1986. However, in a prophetic statement in his letter, Townsley finished, "It must be recognized that our intent to remove all facilities from Fishing Bridge must be politically and socially accepted."[7]

In the interim, those political factors became apparent. In a 1983 grizzly bear recovery plan and again in a 1984 Fishing Bridge report, Yellowstone managers reiterated their intentions to remove the facilities as planned. NPS Director William Penn Mott endorsed removal of the developments. Environmental groups supported the proposal, eventually taking the case to court, arguing unsuccessfully that Fishing Bridge violated protection of the grizzly under the Endangered Species Act. Meanwhile, however, nearby economic interests and federal politicians also weighed in. Commercial tourist interests in the town of Cody, the gateway community east of Fishing Bridge, argued that closure of the facilities might adversely affect their businesses. Those interests were represented by the Wyoming congressional delegation, which asked the Park Service to review the decision. Two months after that expression of concern, the NPS began preparing a new environmental impact statement for the proposed policies.

In the new procedure the agency considered five different programs of action. At the time, Fishing Bridge contained the campsite, the RV park, a visitor center, a gas station, and a large-volume convenience store. The options ranged from removing some campsites but retaining all support facilities to removing all facilities. The last alternative, the plan called for in 1974, was dismissed because historic structures (the visitor center) would be demolished, some campers might have to leave the park to find sites, concessionaire sales would decrease by 7 to 9 percent, and RV concession income in the park would be cut by 13 to 15 percent. The final plan combined the other options to call for moving camping facilities and the gas station to nearby Lake and Bridge Bay areas but retaining the store, the visitor center, and the RV park. The environmental impact study concluded, "The RV park and some support facilities would remain; these facilities do not currently contribute significantly to bear/human conflicts in the area."[8] When I asked a Yellowstone official how much the Cody interests had affected this decision, he smiled and responded, "What do you think?" When I asked members of the Cody Chamber of Commerce to express their view, they called closure of the campground a "shame," then quickly added, "But the RV park is open." Thus, the more natural form of recreation (camping) was sacrificed, but the store, the mobile homes, and the associated traffic were retained.

Meanwhile, the construction of Grant Village continued. The project included 700 lodging units for up to 2,800 visitors a night, three restaurants, housing for private and public employees, service stations, stores, a new marina on the lake, a sewage system pumping treated refuse into the lake, and a four-lane parkway connecting the village to Grand Teton Park to the south. The projected cost as of 1979 was $40 million. By the late 1970s, rangers, scientists, and accountants were questioning the project's necessity, cost, and effect on grizzly bear habitat. These objections caused the project to be stalled several times only to be reborn, in an interesting precedent for the Steamtown case, because so much money had already been spent that aborting the development would constitute wasted funds. Grant Village construction was stopped one more time in 1981 by, of all people, James Watt's special assistant, Ric Davidge. Davidge considered the project wasteful and better left to free enterprise. Davidge was overruled, however, by pressure from the Wyoming and Montana congressional delegations. Construction resumed.

Because the grizzlies lost both habitat areas and feeding sites at the garbage dumps, the NPS might have altered other policies. The agency formed and participated in interagency study teams to determine how "to recover and maintain a viable population of grizzly bears in the ecosystem."[9] Some members of these teams, other observers, and even Superintendent Townsley began expressing support in the early 1980s for the supplemental feeding idea advocated by the Craigheads years before. These "ecocenters" would provide freshly killed elk—a species that now overpopulates the park because its natural predators have been destroyed—in remote areas to provide nutrition and to keep the bears within the park. In one sense, at least, this program would provide poetic justice because the abundant elk have made great inroads in the natural forage on which bears depend. The potential program was denounced and rejected by political appointees heading the teams and by the influential heads of Washington-based interest groups.

CURRENT SITUATION. Today few grizzlies remain in Yellowstone. Those that have survived recent policy changes continue to roam outside the park with fatal results. Notice the two columns in table 4-1 that show mortality rates during the 1980s. Grizzlies obviously continue to wander over park boundaries in search of food and habitat only to be killed on neighboring lands. Some say that grizzlies may yet make a comeback in the park. Park managers have shut off nearly 20 percent of the backcountry to humans in an attempt to give the grizzlies some protection. Even the optimists, however, concede that the bears' chances depend on maintaining abundant and suitable habitat. Preservation of adequate habitat is a dubious prospect.

Suitable habitat outside the park is fast disappearing as a variety of commercial interests invade the Yellowstone ecosystem. The park is now surrounded by logging and mining operations. When NPS officials attempted recently to address this situation with a planning document, they were sanctioned—and the regional director, Lorraine Mintzmyer, was transferred—by politicians such as Senator Alan Simpson and political appointees like John Sununu. Even if the original planning document had been approved, support outside the park for grizzlies is often difficult to obtain. As the response to recent proposals to reintroduce wolves to Yellowstone shows, ranchers and others are not at all well-disposed toward wandering predators.

Bear habitat inside the park, as the stories of Grant Village and Fishing Bridge reveal, is also disappearing. The NPS has closed the campground at Fishing Bridge, but the other facilities remain. Official documents no longer call for the removal of these operations. The 1990 regional planning document only calls for the "phase out of most overnight use at Fishing Bridge." The 1991 Yellowstone *Statement for Management* explicitly exempts the "358-space RV park" from removal plans.[10] Further, the campsites that have been moved were mainly relocated to nearby Lake and Bridge Bay campgrounds. The continued heavy campground use of these areas, which once were pristine bear habitat, is a tribute to the political clout of local economic interests. Agency data show that even though the eastern entrance (from Cody) services only 18 percent of all entering traffic, eastern campgrounds hold 49 percent of the park's campsites. In 1992, environmental groups demanded a halt to the development of facilities in the area and threatened to sue again on behalf of the embattled grizzly species. For the bears, the action may be too little and too late.

SUMMARY. The bears in Yellowstone, particularly the grizzlies, have been jeopardized by the programs of the NPS. The bear population has been severely reduced by loss of habitat and forced changes in feeding habits without compensatory transition policies. The NPS made mistakes in the past because it did not adequately understand ecological needs. Since rec-

ognizing these mistakes as early as 1974, the NPS has proposed some
redress but has been overruled by higher political interests. Even if the
bears are helped by recent changes, they have already suffered from two
decades of political interference. Some critics, notably Alston Chase, com-
pellingly castigate the agency for its actions. I argue that critics should
consider former director George Hartzog's version: "Mr. Chase is hunting
the wrong rabbits; they are not in Yellowstone. They were in Washington,
and maybe some of them still are."[11] Regardless of who is to blame, the
grizzly, with few other homes in the lower forty-eight states, is truly
endangered.

The fact remains that few will see grizzly bears in Yellowstone. Recently, as I walked through the deserted Fishing Bridge campground, I
could smell the fumes from the nearby gas station, see the overflow parking
at the convenience store, and hear the RVs rumbling by into their park.
Certainly, grizzly bears, having more acute senses than I do, were nowhere
to be seen. In what may be viewed as ex-post justification for previous
policies, NPS documents state "when the grizzly bear management program is working well, visitors do not see bears."[12] I had to acknowledge
that it was working that day.

Why You Will See Plains and Wood Bison in Elk Island

I set my pack down and sat on the trail. I hardly moved for what may
have been hours. With Oster Lake on one side and a steep cliff on the
other, I could only go forward or backward. I could see the clearing up
ahead for the backcountry camp, but in between were the reasons I could
not advance. Three large plains bison stood on the grassy area at the head
of the trail. At my approach, all three had stood up and stared at this
intruder, not menacingly but also not invitingly. Now they seemed to be
working in shifts. One would watch me while the other two grazed.
Finally, just before sunset, they moved enough that I could slide by, one
eye locked on theirs and one eye on my only potential escape route, an
early evening swim in the lake. We all slept in the clearing that night.
When I got up in the morning, they seemed somewhat more resigned to
my presence. Nevertheless, I skipped breakfast and slipped out the rear
end of the campsite. After all, I was in their park, not mine.

Elk Island is a remarkable park. It is remarkable for its abundance of
wildlife even though it is close to Edmonton, one of Canada's largest cities.
It is also somewhat remarkable for its policies toward that wildlife. As in
my situation, people adjust, not the animals.

BACKGROUND.   Elk Island is an island in the symbolic sense of the word.
Located in the Beaver Hills area of Alberta, the land is host to aspen and

mixed-wood forests, lakes, ponds, and wetlands. These diverse natural attributes historically spawned populations of moose, elk, beaver, deer, and coyote. When the elk herds were hunted to near extinction, local residents petitioned the federal government to set aside an elk preserve. Elk Park was created in 1906 and then renamed Elk Island in 1913. The most obvious reason to call it an island is that its 194 square kilometers now compose the only national park in Canada completely surrounded by a fence. Since receiving protected status, the park has become a haven for a variety of plants, animals, and birds. Moose, for example, were re-introduced to the parklands around 1910. Beaver were rereleased in the 1940s.

The park is unique in its harboring of both plains and wood bison. Both species were close to extinction when they arrived at Elk Island. Plains bison are the smaller, stockier version with the rounded head that is more familiar to citizens of both the United States and Canada. The plains bison has a tragic history that has been told many times. The population of this animal in North America was reduced, through hunting and slaughter, from 60 million to only a few hundred by 1880. In 1909, the Canadian government was shipping the world's largest remaining herd to another park when about fifty head escaped. That group grew into Elk Island's herd. Wood bison, the larger and darker species, came even closer to extinction. By 1891, fewer than 300 remained in the world. In an in-credible blunder, Canadian park officials allowed much of this herd to interbreed with plains bison while the herd grew. Fortunately, a small, pure herd was discovered in 1957, from which twenty-three head were shipped to Elk Island. The wood bison have been kept in a fifty-nine-square-kilometer area separated from the plains species by a fence and a roadway.

RECENT ACTIONS.   Under the management of Parks Canada and the CPS, different species of plants and animals have prospered at Elk Island. Park managers are guided, as the chief warden told me, by one overriding strategy—"protect the resource first and then provide for the public." They use a variety of tactics for wildlife management. One tactic is to maintain biological diversity. Simply concentrating on one species might ignore other important links in the food chain and the ecosystem. Further, dif-ferent species contribute to revitalizing different aspects of the ecosystem. The beaver population, for example, creates and maintains the wetlands in the park. Today the park hosts 44 species of mammals and 230 species of birds. The mammals range from North America's largest (the wood bison) to the smallest (the pygmy shrew). The diverse forests and meadows are home to hundreds of plant and tree species. The wetlands in the park are internationally recognized as significant. Park managers assist this diversity by introducing or reintroducing species to the park. For instance, trumpeter swans were returned in the late 1980s and fishers in 1990.

A second tactic involves limiting populations when they get too large. The bison are especially productive and thus risk achieving numbers that might result in disease, starvation, or damage to the vegetation. Surplus bison are made available to other agencies, to zoos, and to the public. The animals are routinely tested for disease and infection. Park managers rely heavily on sophisticated simulation models for determining levels of both vegetation and ungulate populations.

A third tactic is using fire to monitor vegetation in the park. Park officials admit that for too long fire was suppressed in the park, endangering the viability of the aspen habitat. Recently the agency increased its use of controlled burns to reestablish grasslands and to maintain the vegetation that serves as food for bison and other mammals. Fire is applied according to scientific projections.

A fourth tactic is illustrative of the story about the elk at the road crossing mentioned in chapter 3. The park is somewhat blessed in that it is entirely enclosed by a fence. Nevertheless, the park does have internal roads, and dangerous interactions between animals and humans are possible. Rather than fencing these areas, the park concentrates on enforcing slower driving and educating the visitor on how to behave around wild animals. One of the sharpest critics of the CPS applauds Elk Island for its emphasis on natural conditions: "A number of commercial activities in the park have closed, and the park is returning to its original preservational purpose."[13]

Finally, the park uses the latest scientific technology in its resource management programs. The previously mentioned simulation modeling is state of the art. Submodels of components of the ecosystem such as the beaver population or biomass are integrated within Geographic Information Systems software to provide an overall model that can simulate and predict. The model is dynamic in that it can be adjusted for certain variables such as precipitation. The Ecosystem Management Model (EMM) is used for various aspects of management. Managers can determine the effect of this year's snowfall on available forage vegetation, for instance . Elk Island personnel work with private-industry consultants and university researchers on these simulations. The momentum for EMM came from within. The park's ecosystems coordinator and other staff members pushed the project, made the outside contacts, and then promoted it so that it could continue to receive government funding.

CURRENT SITUATION.   Wildlife are thriving in Elk Island. Beaver now number close to 3,000. The bison populations are healthy enough for the park to have shipped both plains and wood bison elsewhere to stock other parks and zoos. The recovery of the wood bison has been particularly dramatic. By 1987, wood bison numbered roughly 2,200 in various parts of Canada. The recovery was complete enough to inspire the Committee on Status of Endangered Wildlife in Canada to downgrade the status of wood bison from "endangered" to "threatened."[14]

Park officials see themselves in the forefront of resource management science. The 1992 planning document calls for a huge commitment to research and scientific management. "Elk Island," the document promises, "will be a centre of research in the management of ecosystems. . . . The park will be known internationally for its outstanding contributions in the fields of resource management and natural history/outdoor education."[15] Such rhetoric could be dismissed as hollow promises were it not for the obvious dedication and interest of the park staff in its mission. As Chief Warden Willman told me proudly, "As little as we are, we play a major role in developing science." Even more important, the park is already obtaining results.

SUMMARY.   You will see wildlife in Elk Island. You will see species that were close to never being seen again. You may see them at closer range than you had planned. Furthermore, the locale is not a zoo or a wild animal park, but rather the natural ecosystem in which wildlife reside. The CPS maintains this "natural" state with high-tech resource science and planning and by enforcing a commitment from visitors to experience the park on the park's terms.

Discussion

Comparison of Elk Island and Yellowstone may seem unfair because Yellowstone is much larger than Elk Island. Elk Island has more visitors per acre (6.46 in 1989) than does Yellowstone (1.24), however. Elk Island is also closer to a major metropolitan area (twenty-eight miles) than is Yellowstone (fifty-three miles). Furthermore, that metropolis (Edmonton) is much larger than any near Yellowstone (Cody). Finally, both parks have been designated as protected areas for more than eighty years. Wildlife has had the opportunity to thrive in both parks, but also the chance to be affected by human interactions. For example, while the buffalo herd in Elk Island is thriving, Yellowstone's buffalo are currently threatened by the disease brucellosis and by fears of the disease among nearby cattle ranchers.

The most obvious comparison for wildlife management is in programs for bison and grizzly bears. Through scientific management, Elk Island's administrators have revived the bison population and removed wood bison from the endangered list. By ignoring scientific management, political authorities overseeing Yellowstone have helped to put grizzlies on the threatened list. In Elk Island, the programs used to preserve the bison have been fostered and implemented by park wardens and officials. In Yellowstone, park rangers and officials have been overruled in favor of visitor services and politicians.

Recreation Control

National parks in both countries are continually subject to new forms of recreation and new variations on old forms. In recent years, American park managers have often encountered high-level support for these new activities when their place in national parks was questioned. The CPS, instead, has moved toward favoring more traditional forms of recreation that involve adaptation to the natural world rather than attempts to alter it.

Why You Can Get Peace but Not Quiet in the Grand Canyon

We had just survived Crystal Rapid. We bounced on our rubber, five-person raft in an eddy just below the boiling whitewater and, in between bailing out cold Colorado River water, congratulated each other on a gusto run in one of the toughest rapids in the world. Suddenly we were aware of a high-pitched whine. A huge pontoon boat loaded down with two dozen sunburned tourists came rolling over Crystal's bottom waves and engaged its outboard motor to resume a hurried journey to the next event on the trip down the Grand Canyon. As the pontoon, or "baloney," boat disappeared down the river, its noisy, motorized echoes continued to reverberate off the mile-high walls of the canyon.

BACKGROUND.   A raft trip through the Grand Canyon on the Colorado may be the most exciting recreation available in the world's national parks. The 226-mile journey from Lee's Ferry to Diamond Creek defies description but inspires nearly all who undertake it. The Colorado River in the canyon has the finest and most abundant whitewater in the world. The canyon itself is stunningly beautiful and nearly as full of history and hiking trails as the river is full of rapids. Shortly after the Civil War, one-armed John Wesley Powell led the first known descent down the river. It took several months in wooden dories. One hundred fourteen years later, three river rats sneaked a similar vessel onto the river at high water and completed a daring, all-out, record journey in thirty-six hours. Today's non-motorized raft trips take about twelve days.

In the interim between Powell's journey and the thirty-six-hour record setter, traffic in the canyon grew, diversified, and came under increasing control by the NPS. About 2,100 people rafted this section of the

Colorado in 1967. Five years later, commercial trips combined with private ones to increase that number to 16,500. The 1972 total alone was more than the total for the entire century following Powell's run. The rapid increase in use not only congested the river and the beach campsites, but also produced problems with garbage and human waste that required solution. In 1973, the NPS froze river use at 1972 levels and initiated a series of scientific studies on Colorado River recreation. Even before the studies were completed, park officials released the 1976 master plan calling for "mitigat[ing] the influences of man's manipulation of the river."[16]

Humans affected the river in many ways. Some of the most consequential were beyond the control of the NPS. Since 1963, levels and flows of river water have been controlled by the Glen Canyon Dam, which is under the jurisdiction of the Bureau of Reclamation. The dam, infamous among river runners, had resulted from a bitter dispute about water use in the west. How the dam was used affected the river in ways not yet understood in the mid-1970s. Three aspects of human intervention in the river experience were subject to NPS controls, however. The first, the volume of visitors and their inevitable effects on canyon resources, inspired calls for quotas. The second and third had to do with how those visitors traveled. Outspoken critic Ed Abbey wrote, "The prevalence of airplanes and helicopters in and above the Grand Canyon is a distracting, irritating nuisance which should no longer be tolerated by anybody. I look forward to the day when all river runners carry, as part of their basic equipment, a light-weight portable anti-aircraft weapon armed with heat-seeking missiles. The same must be said of motorized rafting on the river; it adds nothing and subtracts much from the semi-wilderness experience still available."[17] While Abbey may have been the only one suggesting heat-seeking missiles, he was not alone in recognizing the problems. A 1971 study reported that along the south rim of the canyon, "the drone of aircraft engines could be heard almost continuously." With a quarter million sightseers flying overhead each year and the present use of the river "dominated by motor-powered trips," others called for NPS action.[18]

RECENT ACTIONS.   Park managers have attempted to preserve the quiet in the Grand Canyon. Their efforts have centered on aircraft and motor-driven rafts. On both issues they have encountered powerful opponents.

The NPS has now tried to deal with the problem of aircraft noise for two decades. Aircraft include high-flying jets, fixed-wing airplanes and helicopters operating tours, and military and private aircraft. The visual intrusion and the noise pollution do not end at the top of the canyon. Park officials have admitted that river runners and backcountry users suffer "an acute disturbance" from the noise. The agency has considered among its options stopping all flights over the parks. The 1977 program settled for restricted altitudes over different areas of the park below which aircraft were not permitted to fly.

Even these compromised restrictions were relaxed in the 1980s. By the middle of the decade, 45,000 flights carrying about 300,000 passengers flew over the canyon each year. The flights are not safe, or at least they are less safe than running the rapids. Since 1977, eighty people have been killed in flight accidents, including twenty-five in a mid-air collision near Crystal Rapid in 1986. Only sixteen river runners have been killed in the Grand Canyon since 1889. Park staff, with the endorsement of NPS director Mott, prepared environmental assessments supporting recommendations to ban flights below the canyon rim and to establish flight-free zones to reduce noise. Before Mott could promulgate the recommendations, he received a note from Secretary of the Interior Donald Hodel instructing him not to make the recommendations public. Sources later revealed that on the day Mott had met with Grand Canyon staffers, Hodel had met with representatives of the $45-billion-a-year aircraft tour industry. Hodel subsequently called the NPS proposals "unprofessional" and demanded censure of the staffers responsible. Assistant Secretary of the Interior William Horn stripped the noise abatement requirements and instead submitted mild recommendations for Grand Canyon air flights to the Federal Aviation Administration (FAA), leaving restrictions on noise pollution up to that agency. The FAA offered only slight restrictions, while still allowing tour flights below the rim of the canyon.

Developments involving motorized river running were comparable. Research studies on noise from outboard motors on raft trips led to explicit recommendations from the NPS. The sound levels from outboard motors are listed in table 4-2. As the table shows, decibel levels from the motors are often over ninety, or higher than the sound of a pneumatic drill at fifty feet. The motors contribute to other problems such as oil and gasoline pollution, and they increase the ability of rafts to haul down the river large parties that can damage campsites and ecological communities. Still, the major problem is noise. As one study summarized the problem, "outboard motor noise is a deterrent to normal relaxed conversation that one would expect in such an environment, a safety hazard in raft operation, and a health hazard to the motor boatmen."[19]

Two factors might outweigh the negative aspects of motorized rafting in the canyon. First, passengers might prefer the use of motors, enabling them to get through calm stretches of water faster and with less strain. Table 4-3 presents the results of a survey of passengers from combination trips, those using both motors and oars, as a population of travelers with experience on both kinds of trips. As the table shows, only about 5 percent of respondents preferred the motors. More than 90 percent described the oar trip as providing a better wilderness recreation experience—a more

TABLE 4-2. *Comparisons of Grand Canyon Sound Levels*

Sources of sound in Grand Canyon	Decibels	Comparative sources
	130	Jet takeoff (200′)
	120	
	110	Riveting machine
	100	Textile weaving plant
40-hp motor near boatman	90	Subway train (20′)
20-hp motor near boatman		Diesel truck (50′)
	80	Pneumatic drill (50′)
	70	Freight train (100′)
40-hp motor at front of boat		
20-hp motor at front of boat	60	Auto traffic near freeway
Windy day in Grand Canyon	50	Light traffic (300′)
Background in Marble Canyon	40	Chicago residential areas

Source: Thompson, Rogers, and Borden (1974, p. 9).

relaxed pace, a more sensitive environment, and more comfortable social groupings. A second factor that could be in favor of motors is economic. Perhaps motorized rafts enable greater profits for the commercial companies running trips through Grand Canyon. Another independent study in the mid-1970s showed that while revenue to the concessioner for oar trips was about $2 less per user day than for motor trips, the cost to the concessioners was $1.50 less for oar trips. The study concluded that "conversion to partial or all oar trips would not place a financial burden on existing concessioners particularly if a two to three year conversion period were allowed."[20]

The NPS used these studies in its 1979 *Colorado River Management Plan*. The plan provided a timetable for the removal of motorized watercraft from the Grand Canyon. The phasing-out period used months in which motors would not be allowed. For example, April and September of 1981 would be quiet months, then April, August, and September of 1982, and so on. By 1985, only oar-powered rafts were to be allowed on the Colorado River in the Grand Canyon. Interestingly, this plan had a legal precedent had it been challenged in court. In his study of the boundary waters, Gladden describes the conflict over motorized boating in areas including Voyageurs National Park. That dispute inspired the court case *United States* v. *Brown* in which the courts supported, and upheld on appeal, the right of the federal government to restrict on national parklands activities, such as the use of motorized vehicles for hunting, not mandated in congressional establishment of the parks.[21] The Colorado River plan was challenged, however, in a political rather than in a legal forum.

The political reaction to the plan was immediately apparent. One year after Yellowstone issued its 1979 *Final Environmental Statement of Proposed*

TABLE 4-3. *Survey of Grand Canyon Travelers*
Percent; actual numbers in parentheses

Question	Oar	Motor	No difference	Total
If you were planning a trip on another river, which type of trip would you choose?	87 (46)	4 (2)	9 (5)	100 (53)
Which would you recommend to a friend planning a Grand Canyon trip?	79 (42)	6 (3)	15 (8)	100 (53)
Which type of trip better enabled you to "experience" the Grand Canyon?	91 (50)	5 (3)	4 (2)	100 (55)
Overall, which type of trip did you like better?	82 (45)	5 (3)	13 (7)	100 (55)

Source: Shelby and Nielsen (1976, p. 16).

Colorado River Management Plan, Congress responded with the Hatch amendment to the DOI appropriations bill. The amendment centered on a section of the U.S. Code calling for "protection of concessioner's investment." The section is intended to "assure the concessioner of adequate protection against loss of investment."[22] The amendment was named for its sponsor, the republican senator from Utah (Orrin Hatch), whose constituents included many, if not most, of the rafting companies that run commercial trips on the Colorado. The amendment effectively guaranteed commercial rafters continued motorized use of the river and no net reduction of summer season traffic in the canyon below 1978 levels. This intervention by higher authorities would be just the first of several into NPS attempts to preserve the natural conditions of the river experience. The NPS later reported, "Due to the controversial nature of this plan and congressional action that limited the agency's ability to implement the approved, preferred alternative, a revised plan was released in 1981."[23] In the revised plan, motors were still allowed on rafts in the Grand Canyon.

To be accurate, the continued presence of motorized rafts and air travel in the canyon did not displease all visitors. Early in his tenure as secretary of the interior, James Watt described his trip on the Colorado River: "The first day was spectacular . . . the second day started to get a little tedious, but the third day I wanted bigger motors to move that raft out. There is no way you could get me on an oar-powered raft on that river—I'll guarantee you that. On the fourth day we were praying for helicopters and they came."[24] With Watt and his disciples in charge of the Department of the Interior for years to come, resurrection of the earlier NPS plans for the river was unlikely.

CURRENT SITUATION. The results of political intervention on behalf of the tour flight operators and motorized commercial rafters is evident in the 1989 *Colorado River Management Plan.* The plan recommends, "Offer, to the extent possible, a primitive river experience without intrusion from aircraft."[25] But the specific provisions are heavily compromised and still require the NPS to answer to the FAA. By last count, flightseeing now involves 400,000 tourists a year in more than 100,000 flights buzzing canyon users roughly every five minutes on all but a forty-five-mile stretch of FAA-cleared air space.

Rafting guidelines limit the total number of user-days on the river to 169,950, with more than 115,000 of those allocated to commercial companies. The plan specifies the number of launches a day in each month, the carrying capacities for different types of rafts, limits on trip size and miles covered a day, and even allocations for individual companies. The plan explicitly allows motorized craft except between September 16 and December 15. Interestingly, in the section of the document discussing public comment, the NPS admits that it still receives questions regarding motorized use of the river. The official response was, "The motors issue was not evaluated." When I asked Grand Canyon's chief ranger about the continued use of motorized rafts on the river, he admitted that there had been tension but added curtly, "The agency stands by the decisions of 1981."

SUMMARY. The Grand Canyon faces numerous threats. Several may be more consequential in the long run than the noise faced by river runners and hikers in the canyon. The NPS must deal with the impact of fluctuating water releases from Glen Canyon Dam, air pollution from power sources on the nearby Kaiparowits Plateau, development proposals on the North Rim, and other issues. The agency's experience with noise pollution should, however, provide some lessons for field managers. The final decisions will not be up to them. NPS managers should also remember that, even with scientific backing, political clout can be a difficult opponent.

Why You Can Get Peace and Quiet in Jasper

I was trudging along through Watchtower Basin near the Maligne River. The recent rains had left the valley along the river marshy in spots, worse in others. The little trail was hard to follow anyway, but with dusk coming on, it was getting to be impossible. I had already seen bighorn sheep and porcupines along the trail, but no sign of other humans, let alone trail markers. The trail guide said I would have to ford the river where I could see a sign on the other side to get to the backcountry camp, but I was beginning to look for other potential camping spots. The only sounds I could hear were the river rolling by and the mosquitoes complaining about my insect repellent. I finally took a chance. Where the trail had disappeared but the river looked fordable, I waded in, and there it was, a tiny wooden marker pointing to the campsite. I had to admit, the CPS was obviously keeping the basin as natural as possible. No surprise there: Jasper is almost heaven for hikers and backpackers who love wilderness conditions.

BACKGROUND.   The largest of Canada's Rocky Mountain parks, Jasper has been hiked by Indians, explorers, and fur traders since at least the seventeenth century. Fur trader David Thompson trapped Jasper's rivers about the same time that John Colter wandered into Yellowstone to the south. By 1907, enough Canadians had seen the rugged mountains, alpine meadows, towering waterfalls, clear lakes, fast rivers, and slow glaciers to support the establishment of Jasper as a national park. Trains carried tourists the 220 miles in from Edmonton to the east. The road from Banff in the south was built during the Depression. Finally, the magnificent Icefields Parkway was completed in the early 1960s. The town site at Jasper grew to about 4,000 residents. Still, when visitors arrived, they hiked. Jasper has always been a hiker's paradise.

By the late 1970s, Jasper's popularity was growing. Not only had visitation increased dramatically, but the visitors were looking for increasingly diverse forms of recreation. Cross-country skis had become affordable. Canoeing and kayaking had become popular. Lightweight gear made backpacking and mountaineering more accessible than it had previously been. Surveys also showed, however, that the public wanted more recreational opportunities, but without damage to the natural environment. The managers of Jasper have obliged.

RECENT ACTIONS.   The first response by CPS managers involved zoning. As described in previous chapters, park officials divide parklands into zones for different categories of use. The zones are somewhat more restrictive in Canada than in the United States. Even using the zones as comparable, however, the amount of land set aside for wilderness in Jasper is impressive. Zone 2 or wilderness constitutes 98 percent of the park area. Contrast that to other large, popular, western national parks. Yosemite's backcountry constitutes 95 percent of the park. Yellowstone's natural zone is 90 percent of the park area. Even Grand Canyon, with its huge expanse of area that cannot be developed, has designated less (97 percent) of its land as backcountry. Approximately 93 percent of neighboring Banff is designated as wilderness. As zone 2 land, 98 percent of Jasper is off limits to motorized access.

TABLE 4-4. *Revenue Value of Building Permits, Fiscal Years 1985–87*
Thousands of dollars

	Residential		Commercial		Total	
Year	Banff	Jasper	Banff	Jasper	Banff	Jasper
1985–86	951	236	6,581	515	19,136	751
1986–87	3,913	326	40,661	861	49,860	1,187
1987–88	8,699	484	30,775	1,005	46,829	1,489

Source: Canadian Parks Service (1989b, pp. 5, 9).

To protect these wilderness areas, Jasper managers must prevent urban sprawl and spillover from the town site and the rest of the zone 5 area. In this regard, the pressures facing Jasper are more comparable to those in Banff than those in Grand Canyon. Jasper's development potential is like Banff's, but on a somewhat smaller scale. Jasper has roughly half as many permanent residents as does Banff. CPS expenditures in Jasper are about half those at Banff. Jasper entertains approximately one-third as many annual visitors as does Banff. In fact, visitor totals actually increased at a faster rate in Jasper than in Banff during the mid-to-late 1980s. As Superintendent Gaby Fortin said in an interview, "There's nothing in the rules stopping Jasper from becoming another Banff."

However, Jasper managers have been able to restrict the growth of their town site to a much greater extent than have their Banff colleagues. Table 4-4 compares the granting of building permits in the two towns over the most recent years for which complete data are available. As the data show, the value of building permits granted in Jasper is magnitudes smaller, in any category or year, than those in Banff. The differences are much greater than can be explained simply by the larger scale of pressures facing Banff. Fortin and the other CPS managers of Jasper have made conscious decisions—by, for instance, limiting permits and restricting building height—to prevent the overdevelopment that they say has occurred in Banff.

The zone 2 land surrounding the town site is not locked up or inaccessible. In fact, Jasper contains more than 1,000 kilometers of backcountry trails. Use of these trails is strictly controlled, however. Park managers have determined the carrying capacity of each trail, that is, the level of use beyond which the natural environment and the wilderness experience would be impaired. The management plan states explicitly, "Backcountry use quotas and other management practices will be used to maintain the quality of visitor experiences and to protect the backcountry environment."[26] Permits to use the trails can be reserved for up to 35 percent of a trail's carrying capacity. The rest are issued on a first-come, first-served basis at the visitor center. Groups are limited to ten persons to reduce crowding, and camping is prohibited except in designated areas.

The natural state of the hiking areas is preserved through several other programs. Wildlife is abundant, one major reason being the program to control poaching. As chapter 3 mentioned, poaching has historically been a threat to Canadian national parks. Jasper officials have recently implemented a five-year program using DNA samples and photographs of elk and bighorn sheep so that wardens can trace hunted animals. One observer described the program, "Jasper National Park wardens are breaking ground in North America."[27] Second, park managers have designed special plans for areas that have become particularly popular with hikers. In 1991, for example, the CPS developed a concept plan for the Maligne Valley that monitors use of the area to ensure its natural condition. The plan states: "Wilderness values will be the foundation of all management actions."[28] Finally, in direct contrast to the motors in Grand Canyon, Jasper now allows rafting, canoeing, and kayaking on its rivers but explicitly prohibits motorized use to "avoid the intrusion . . . into backcountry areas."[29]

CURRENT SITUATION.   Jasper's backcountry remains an expanse of rugged country with peace and quiet for those who use it. Some express skepticism that the CPS will keep it this way in the face of growing visitation and commercial pressure. They cite the appearance of horse outfitters and whitewater rafters as evidence of increasing use of the backcountry. I asked Superintendent Fortin about park supporters' fears of accelerated development. "Their concerns are legitimate," he responded, "but we aren't gonna let that happen." Obviously, talk is cheap, but the park's recent actions support his statement. Furthermore, he and the other CPS staff are supported by nearly all other interested parties. Local Chamber of Commerce President George Andrew admitted to differences of opinion with Fortin but added, "Look, I spend a lot of time in the park. I don't want to see it screwed up either." Perhaps more important, CPS field officers like Fortin are backed by their own higher officials. As mentioned in chapter 3, regional officials recently made park interpreters and information staff heritage communicators. The Jasper staff have taken to the new name and mission. The park's own brochure states, "With the new name comes a stronger mandate to promote environmental stewardship and preserve the park's ecological integrity."

SUMMARY.   No national park is a perfect wilderness. Jasper has its share of challenges, not the least of which is preventing excessive growth of its town site. Get out into the park, however, and the effects of the CPS orientation toward preservation become apparent. More accurately, perhaps, human intrusions into the natural state are noticeable mainly by their absence.

Discussion

This comparison between Grand Canyon and Jasper concerns recreation in natural spaces. Both parks are known for awesome scenic vistas and challenging backcountry recreation. Jasper is twice as large as Grand Canyon and serves only one-third as many annual visitors. One might suggest, therefore, that maintaining a pristine condition is easier for the less-crowded park. The discussed contrast, however, was not based on sheer numbers. Both Grand Canyon and Jasper limit the use of their backcountry resources with quotas on river and trail travel. The contrast is more marked in two respects. First, one park has not been able to prohibit noisy intrusions—motorized rafts or urban sprawl—into backcountry areas, while the other has. Second, one park's decisions have been countermanded by higher political levels, while the other's have been supported.

Commercial Development

Both park services must manage commercial developments within the parks. These developments have resulted from the need to provide essential services to visitors but also from decades of providing amenities and entertainment as well. Recent efforts in Yosemite and Banff display different patterns of behavior by the agencies but, at this point, unclear results.

Why You Still Get Stuck in Traffic Jams in Yosemite Valley

My workdays in Yosemite began with an early morning stroll from employee housing across the valley to the office of Grounds and Maintenance. Even in midsummer, the mornings were cool and quiet, with the pine smell heavy in the stillness. The mountain air was clear, and the view of the sun coming up over Half Dome to the east was inspirational. I would actually get excited about the day ahead, even if it did involve scrubbing outhouses. By later in the day, however, the view of Half Dome was obscured by the smoke from campfires in Lower Pines and North Pines campgrounds. Crossing the valley meant dodging traffic on the park's loop road. The quiet and the pine fragrance were both obliterated by vehicle noise and fumes. On paper, the park had instituted actions to change this scenario since I worked there in 1977. I went back in 1992 to see what had been accomplished.

BACKGROUND.   Yosemite is, for many, the prettiest place in the United States. The park comprises more than 700,000 acres of mountain valleys, beautiful waterfalls, lush meadows, and big, in all senses of the word, forests. John Muir, the explorer whose name is almost synonymous with Yosemite, wrote, "But no temple made with hands can compare to Yosemite. Every rock in its walls seems to glow with life."[30] The heart of the park is Yosemite Valley, the home of such natural wonders as El Capitan, Yosemite Falls, and Mirror Lake. As Muir said, "The most famous and accessible of these cañon valleys, and also the one that presents their most striking and sublime features on the grandest scale, is the Yosemite."[31]

The mile-wide, seven-mile-long valley has also been at the heart of controversy for decades. Muir's comment portends inevitable tension in that the valley is both scenic and accessible. Even before cars were allowed in the valley, British ambassador Lord James Bryce offered a prescient warning: "If Adam had known what harm the serpent was going to work, he would have tried to prevent him from finding lodgment in Eden; and if you were to realize what the result of the automobile will be in that wonderful, that incomparable Valley, you will keep it out."[32] Instead, during the 1930s, California built the All-Year Highway along the Merced River, allowing automobiles to enter the valley from the west. In the following decades, thousands of visitors poured in. By 1954 a million people a year visited Yosemite, a vast majority entering through and often staying in the valley. Concessioners and developers responded with stores, hotels, and other man-made attractions in the valley. As historian Al Runte wrote, "Yosemite Valley inevitably became the symbol of the national park both at its finest and at its worst."[33] The valley was home to such questionable activities as the famous firefall, an event in which burning wood was pushed over the 3,500-foot Glacier Point cliff in the evening to provide a spectacle to those watching below. By 1970, Yosemite was hosting more than 2,270,000 visits a year, entertaining most of the visitors in the valley.

Questions about NPS management of the valley in the 1970s were stimulated by many factors. The sheer volume of visitors had become problematic, with traffic jams and automobile pollution on summer days. A counterculture celebration in Stoneman Meadow turned violent. The new owner of concessions in the park, MCA, announced plans for a television series in the valley, more development, and a tramway to Glacier Point. Park managers themselves came under fire for proposals both to remove auto traffic from the valley and to endorse the tram project. In 1974, the park's master plan was rejected and a new planning process opened for future management of the park and the valley.

RECENT ACTIONS.   The NPS revealed its plans for Yosemite with the 1980 *General Management Plan* (GMP). The plan resulted from years of public comment and open discussion involving more than 60,000 people and more than forty public workshops. The 1980 GMP was to provide a "comprehensive set of decisions regarding the future of Yosemite." In the plan, Yosemite's managers admitted that because of past actions, "Today, the Valley is congested with more than a thousand buildings . . . [and] approximately 30 miles of roadway which now accommodate a million cars, trucks, and buses a year." The GMP then explicitly stated, "The intent of the NPS is to remove all automobiles from Yosemite Valley and Mariposa Grove and to redirect development to the periphery of the park and beyond." The plan backed up these general preferences with specific recommendations for overnight and day use levels, expanded shuttle bus services, redesign of Yosemite village, and removal of numerous buildings, activities, and 1,000 parking spaces. Many operations, including park administration and housing, were to be moved to the little town of El Portal, just west of the valley. While it did not provide an exact timetable, the GMP did use "the next 10 years" as its framework.[34]

Over the next dozen years, little was done to put the plan into effect. Shuttle bus service was expanded, more than 300 parking spaces were eliminated, and some rangers' quarters were moved to El Portal, but few other specifics were accomplished. By 1984, Superintendent Robert Binnewies had already implicitly accepted a compromise goal: "I have worked very hard on the theme that a certain level of development has been achieved within the park and that that level should now be held and no additional expansion or development in the park should occur."[35] The comment acknowledged that reduction below present levels was unlikely. Referring to the 1980 GMP in 1986, a project design team involving regional and park officials concurred, concluding, "In the foreseeable future, however, it is doubtful that implementation could occur."[36] By the end of the decade, James Ridenour, director of the National Park Service, was calling the plan only "a concept, a good ideal."[37]

What happened instead truly represents the conversion of preservation ideals to utilitarian outcomes. Table 4-5 uses data provided by the NPS and the Wilderness Society in the 1990 congressional hearings on agency concessions policies. The table shows the expansion that has occurred in the valley in spite of the 1980 GMP and the desires of park managers like Binnewies. Rather than returning to a more natural condition, the valley has been even further developed by commercial interests. I asked park historian Alfred Runte if the valley had been naturalized at all by the 1980 GMP. "No," he answered; "business as usual."

Why? Political intervention in park actions may have been less overt in Yosemite than in the Grand Canyon and Yellowstone situations, but it was still significant. In this case, Congress prevented implementation of the plan simply by not providing the money. The 1986 project study esti-

TABLE 4-5. *Yosemite Valley Development in the 1980s*

New concessioners[a]	Expanded concessioners[b]
Camp Curry pizza and ice cream stand	Marketing, catering, conferences
Raft rental	Bicycle rentals
Wine tasting, gourmet, and music festivals	Cross-country ski rental
Photo finishing center	Expanded retail space, restaurant, and bar
Video rental	at Yosemite Lodge
Tobacco shop	Breakfast service at village store hamburger
Camp store at the ice rink	stand
Tour sales outlet at village store	
Fudge counter at village store	
Moonlight tram tours	

Source: U.S Congress, Senate Subcommittee on Public Lands, National Parks and Forests Hearings (1990, p. 110).

a. Concessioner profit centers established since 1980, not in the General Management Plan.

b. Existing concessioner centers expanded since 1980.

mated costs for the 1980 plan of at least $22 million for construction of parking areas outside the park, $33 million for new buses, and $19 million for annual operations. Superintendent Mike Finley, in an interview, explained the decision as purely financial and not political. "Our annual budget is only 15 million dollars here," he said, "and that project ran over seventy." Furthermore, the Yosemite budget now is roughly the same as it was ten years ago. Of course, budget decisions are political calculations, as the $63 million estimated final cost for Steamtown, described in chapter 2, exemplifies particularly well. In the management plan process, all levels of government are involved and presumably aware of potential costs for plans. Thus authorities at all levels committed to the Yosemite GMP in the 1980 planning process. Had the Steamtown money been appropriated for Yosemite revisions as called for in the 1980 plan, the valley would look much different today.

Congress did not appropriate the money, in large part, because of local economic opposition. The political clout of the concessioners operating in the valley has become quite apparent in recent negotiations over a revised GMP. In 1989, Yosemite Park & Curry Co. began an intensive campaign to get even more development in the park. The company sent out 93,000 letters to former guests to support efforts to abandon the 1980 plan. The concessioners are strongly supported by many members of Congress. In 1990 hearings, Secretary of the Interior Manuel Lujan, Jr., made the mistake of criticizing the current low fees charged to concessioners, citing particularly the Yosemite situation. Senator Malcolm Wallop (R-Wy) responded sharply, "Secretary Lujan's appraisal of the situation is a completely ignorant view of ordinary business practice in America."[38] Furthermore, the park's concessioners had the support of the NPS director, James Ridenour. Ridenour referred to Yosemite Park & Curry Co. as a viable contender for the new park concessioner contract. Well aware of the 1980s expansion counter to the GMP, Ridenour said, "By all measures we generally take, their performance is pretty good."[39]

CURRENT SITUATION.   Indeed, a new contract was recently awarded in Yosemite. Economic measures suggest that the account should be quite desirable. In 1989, more than 3.4 million people visited the park. In the same year, Yosemite Park & Curry Co. had gross revenues of $85 million, while it only paid $636,000 in franchise fees. In early 1991, MCA, the owner of Yosemite Park & Curry Co., sold its possessory interest temporarily to the nonprofit National Park Foundation. The NPS then took bids from several possible buyers including the incumbent, Disney, Marriott, and the Yosemite Restoration Trust. As chapter 2 described, the contract was awarded in late 1992 to a Buffalo-based sports concessioner. While the new contract did increase the financial return to the park from concessions operations, the ultimate impact on the valley is not clear.

Current attempts by Yosemite managers to revise the GMP reflect tensions that will be involved in the new concessioner operations. In late 1991, officials released a new plan calling for a 13 percent reduction in overnight accommodations in the valley and removal of some tennis courts and the park's ice rink. This plan also called, however, for conversion of rustic tent cabin lodgings to higher-priced motel units and the expansion of fast food facilities in the valley. In 1992 public hearings, this plan stirred considerable opposition. Environmental groups condemned the proposal for ignoring traffic problems and continuing to turn the park into a resort. These participants instead called for a reconsideration of the 1980 Plan.

SUMMARY.   When a friend and I returned to Yosemite fifteen years after I had worked there, I found little had changed. The park was still incredibly beautiful, but also disturbingly crowded and congested. Current park managers had imposed new limits in the summer of 1992, banning morning and afternoon campfires in the valley. Still, by early evening a low haze hung over the eastern end of the valley, like a cloud below Half Dome. I realized that the 1980 plan had not changed much. By now its goals of reducing human impacts were officially abandoned. I thought about a recent statement by NPS Director Ridenour on the current plan for the future of Yosemite: "We based it on the [current] level of commercial activity, with some growth just in terms of inflation. We essentially are saying the level of commercial activity is at its peak. It's not going to get any bigger."[40] This vow of "preservation" at current development levels reminded me of the GMP promises of 1980, the Binnewies comment of

1984, and so many other calls to "preserve" the valley. Fortunately, I realized, Tuolumne Meadows and the relatively undeveloped high country of the park have not received the same "preservation." We decided we could make Tuolumne by dark. We loaded up my van and headed out of the valley as the oncoming traffic rolled in.

Why You Still Get Stuck in Traffic Jams in Banff

Half of the visitor center in the town site of Banff looks like many other national park visitor centers. The visitor can find weather reports, maps, helpful park employees, and even a display or two. The other half of the center reveals something basic about Banff that is unique in its excess. The other half of the same room housing the CPS booths is occupied by the Banff Chamber of Commerce. A visitor can make reservations for a campsite at one booth, do an about-face, walk twenty feet, and reserve a hotel room for the next night. While this may seem convenient, it also represents a potential contradiction for the CPS that has yet to be resolved.

BACKGROUND.   Banff National Park is Canada's pride and joy. Far and away the most famous and most visited of Canada's parks, Banff is internationally renowned for its mountain vistas, glaciers, hot springs, and lakes. Many consider Lake Louise the prettiest single spot in the world, while a replica of neighboring Moraine Lake adorns Canadian currency. Banff has been a popular tourist destination since the transcontinental railroad made visits to the hot springs easy. The railroad companies constructed the first building at Banff, the Banff Springs Hotel, in 1887, and the town grew up around it. The park's first superintendent prematurely sold town lots instead of leasing them after a telegram ordering that lots be leased but not sold was garbled as "leased or sold."[41] It was only the first questionable decision that allowed the town to grow into today's small city.

The town site's growth was not seriously questioned until much later in the twentieth century. Once completion of the Trans-Canada highway from Calgary to Banff in 1911 made the park accessible to more than just railroad passengers, visitors and developers moved in. Over the next fifty years, the leasing system described in chapter 3, involving long terms, low fees, and nearly perpetual renewal, allowed and indeed encouraged Banff town site to grow. After several public disputes, Parks Canada let ski resorts develop so that the park could become a year-round destination. Park agencies also helped by building and managing infrastructure such as sewage systems and roads. By the end of the 1960s, the town site had expanded to the point that some park scholars began to question the government's pro-development policies. Nevertheless, in the following decade, most Canadians applauded the town site and its residents. One visitor wrote in 1975, "Today it is a Canadian town, and one in which all Canadians take pride." Another observer of the late 1970s described, but did not criticize, specific questionable aspects of the town site, such as the airfield. Still another defended the leasing system and characterized the fees as higher than was generally understood.[42]

One simple explanation of why development of the town site was fostered and encouraged by the park agencies involves the concept of local capture. Park officials not only worked in Banff but also lived in the town site. They, too, were residents. Their neighbors ran local businesses. Their children went to school with the children of other residents. Any relationship between park managers and town residents other than mutual sympathy and support would have been surprising. One piece of evidence for this compatible relationship was the lukewarm support of residents for independence from the park agencies. A 1970 plebiscite showed that a majority of residents favored local self-government, but there was a low turnout for the vote. Most residents did not find agency management problematic.

RECENT ACTIONS.   The relationship between park officials and residents started to change in 1979. Several policy changes in that year were directed at town site residents. First, inexpensive, virtually perpetual leases were mandated to be replaced with limited-tenure contracts, wherein renewal was subject to review. Officials and residents alike predicted that such a change would lead to large increases in Banff rents. Second, Parks Canada promulgated the "need to reside" policy mentioned in chapter 3, limiting residency to providers of essential services. In Banff, however, the policy was potentially limited in that it was directed only at property purchases and not at existing renters. Third, the agency stated pointedly that, in the future, towns were "to be contained within boundaries as set in legislation."[43]

While these policy changes were mandated systemwide, the changing attitude toward town sites was particularly acute and noticeable in Banff. The changes were endorsed through the ranks there and implemented in subsequent actions. A 1979 editorial by Banff warden Ronald Seale criticized the town site and praised the absence of similar intrusions on natural settings in American parks. Chief warden Sid Marty called previous accommodations with the Banff town site "bitter but necessary compromises" and blamed them for increasing wildlife problems. Superintendent J. E. Vollmershausen stated explicitly that he would "discourage" businesses trying to set up nonessential attractions in the park.[44] Park officials fought the proposed twinning of the Trans-Canada Highway section in

Banff, an action that drew rare praise from the Canadian Nature Federation: "Perhaps the most startling development was the vigour with which Parks Canada attacked the . . . project and came to the defence of national park values."[45] More recently, park officials quietly amended regulations and temporarily shut down the local airstrip to the great consternation of local residents.

Why did CPS employee attitudes toward town sites change from the previous pro-development stance? As chapter 3 described, the entire agency underwent a certain amount of soul-searching in the 1980s. Whereas CPS employees resigned themselves to some development in other parks with town sites, the historical tendencies to favor development over preservation were exaggerated in Banff. As former superintendent Steve Kun said, Banff needed to "have the lid put on it as far as development goes."[46] The symbiotic relationship with town site residents was increasingly replaced by professional concern for environmental values. As a result, relations between the CPS and residents became "very contentious," according to current superintendent Charlie Zinkan.

Park employees could only go so far in Banff. Whereas delegation to field personnel in other Canadian parks continued to increase during the decade, Banff managers were occasionally reined in. In the airstrip situation, for example, angry constituents called on local MPs Joe Clark and Albert Cooper to respond. The MPs were able to persuade Environment minister Jean Charest to back down and reopen the airstrip. A second example of curtailed discretion for Banff managers occurred in 1989 when opponents of a proposed expansion of one of the park's ski resorts called on the government to "unmuzzle" park officials.[47] The political clout of Banff residents is legendary among CPS employees. Eighteen-year Banff veteran Perry Jacobson told me, "If you did something someone there didn't like, you could bet you'd be hearing from Ottawa a half hour later."[48] Superintendent Ian Church of neighboring Yoho agreed: "The old joke was, if the super turned down your building permit, you could just fly to Ottawa and get the MP to approve it." Certainly the volume of permits reflected in table 4-4 lends some credence to Church's comment.

The result has been a rather puzzling mix of programs in Banff. While field employees were battling commercial interests over the highway, airstrip, and ski resorts, a Parks Canada report in 1984 called for a 30 percent increase in hotel rooms in the town site in spite of public comments running heavily against new construction. The management plan for the park released in 1988 is also somewhat ambiguous. The plan explicitly "contains recognition that limits to the future growth and development of the Banff town site region must be established." In the specific section on the town site, however, the document admits, "Although the original intention had been to do so, the park management plan does

TABLE 4-6. *CPS Revenue in Banff, Selected Categories, Fiscal Years 1987–90 Canadian dollars*

Service	1987–88	1988–89	1989–90	1990–91
Sewage	162,576	65,313	50,122	9,543
Water	533,103	281,080	216,950	23,070
Garbage	419,061	379,092	307,006	93,357

Source: Canadian Parks Service (1991c, p. 70).

not include a detailed town boundary due to the complexity of the issue and time needed to make the final determination."[49] Through the 1980s, the town continued to grow.

CURRENT SITUATION.   Banff town site is now the city of Banff. The disputes with park officials during the 1980s gave residents who had long sought self-government enough momentum to obtain the necessary political support. A federal-provincial agreement in 1990 recognized Banff as a municipality with the power to levy taxes and the responsibility to provide services. The CPS no longer manages functions like fire protection and sewers within city limits. In fact, table 4-6 shows how the CPS has weaned the city from certain services over the past few years. The table displays agency revenue from sewage, water, and garbage services within the park. Some amounts remain in the 1990–91 column, even after the town site became a recognized city, because the CPS still provides services in other parts of the park, such as the ski resort at Lake Louise.

The city council has a difficult job. Banff has more than 7,000 permanent residents and more than 10,000 units for overnight visitors. One of the first crises with which the city council will have to deal is resentment by many residents toward increasing Japanese ownership of hotels and restaurants. Whether investment comes from foreign sources or not, however, the pressures for greater commercial development will remain strong.

What this new autonomy for the city means for preservation efforts is not yet clear. Superintendent Zinkan admits that many feared the change would result in a more pro-business orientation but adds that the opposite has so far been true. The city council has been more restrictive than the CPS could have been, according to Zinkan, because residents now realize that "they have to take care of their own problems."[50] Park officials have established a good working relationship with some of the most prominent businesses. For example, Brewster Tours, the company responsible for the huge tour buses rumbling up and down the Icefields Parkway, now claims to incorporate CPS strategies of environmental education into its mission. Yoho's Church agreed that the city had so far done a "good" job, but he added that his "gut feeling" told him it was wrong. "It won't be long,"

Church said, "before they [Banff] panic over losing market share to As-pen." Kootenay's chief warden, Perry Jacobson, was also skeptical. He was working in Banff when the town was officially declared but said he was "never convinced that making it autonomous was a good move." Like the first message in to Banff, the latest message coming out is garbled.

SUMMARY.   Banff is the greatest remaining anomaly in the CPS shift to preservation. Other parks, as chapter 3 described, have moved to adapt their town sites to the natural setting. Banff is different, as one observer noted: "What makes Banff's dilemma so extreme is that pressures on the town have been for development, while management of the rest of the park has become increasingly conservation-oriented."[51] In the past two decades, the relationship between the town and the park management has gone from mutually sympathetic to antagonistic to independent. Whether or not the transfer of power to the town was another case of the CPS being smart is unknown because the consequences are not yet clear.

Discussion

Yosemite Valley and the city of Banff are closely comparable. Both are pockets of civilization surrounded by awesome natural splendor. Both are world-renowned vacation destinations. Banff entertains roughly 4 mil-lion visitors a year, while Yosemite hosts about 3.5 million. Banff has more permanent residents and hotel units, but the Yosemite Valley may have more vehicle traffic. Both parks have experienced foreign investment and resentment toward that investment. Both parks have also witnessed a shift in attitudes among those who manage them toward favoring natural preservation. And, in both cases, that shift has been countered by the political clout of pro-development interests. The future direction of each park remains undecided.

External Threats

The NPS and the CPS are both increasingly aware of external threats facing the parks. Both have developed research programs to study and provide information necessary for addressing those threats. The experi-ences of the services have differed, as the cases of Shenandoah and Prince Albert show, in how that research is received.

Why You Cannot See the Forest or the Trees in Shenandoah

Like thousands of others, I made my latest trip to Shenandoah in the month of October. Fully 20 percent of the park's 2 million annual visits occur in October, more than any other month. The reason became obvious as I drove west through the Virginia countryside. The orange and yellow leaves of oak and birch trees fluttered in the autumn breezes as I approached the park. Like my fellow travelers, I anticipated the views from Skyline Drive and checked again to make sure I had remembered my camera. When I finally did get to the scenic overlook at Big Meadows Lodge, I was stunned, but not by the fall colors. Nor was I alone. Looking west, a young man standing near me wondered aloud, "What is that, smog?"

BACKGROUND.   The views from the Blue Ridge skyline were a major reason for the park's establishment. In its review of the park's history, the management plan states, "Outstanding scenery is a key to the identity of Shenandoah, as the views from the ridge were a primary consideration in the creation of the park and the Skyline Drive." The park was actually "re-created," as Franklin Roosevelt said when he dedicated the Skyline Drive. The Shenandoah valley was well settled by the middle of the nineteenth century. By 1900, "virtually every piece of land that could produce anything was used."[52] The mountain economy worsened, however, partly as a result of a chestnut blight that destroyed many of the local trees in 1915. During the same time, mountain resorts became popular. In 1923, NPS director Stephen Mather urged the establishment of a park in the Appalachian range. Following a commission's recommendations, Shenandoah was designated in 1926. All land for the park had to be secured only by public or private donations, and the state of Virginia began to purchase private lands from local mountaineers, lands that it donated to the park. Skyline Drive, a road running the eighty-mile length of the park along the ridge tops, was built as a public works project during the 1930s. The original park area, re-created from private property with the help of the state of Virginia, provides the basis for an ironic twist decades later.

The park built a reputation for scenic vistas that was unchallenged well into the 1980s. In 1977 the NPS sponsored a symposium to provide research assistance to the park. The papers prepared for the symposium covered subjects such as critical habitats, forest types, wildlife populations, and the effects of fire. The papers did not even consider air pollution. Even the 1983 management plan does not address air quality. The plan instead identifies "four general categories of long views or vistas that are gained from the Skyline Drive and its overlooks" and even provides a map of the scenic panoramas.[53]

Though scenic vistas at Shenandoah and other parks were not yet impaired, NPS attempts to prevent future problems had already been overruled. In 1981, the agency identified 173 such "integral vistas" in forty-three national parks, including Shenandoah. Identification was made explicit to allow states, acting under the Clean Air Act, to recognize potential harm to such vistas when considering proposed new developments outside park borders. In order to be considered, the list had to be sent from the NPS to the Environmental Protection Agency (EPA) by December 31, 1985. In spite of the efforts of NPS director Mott and other agency officials, Interior secretary Hodel never sent the listings to the EPA. Thus these scenic vistas became even more vulnerable.

RECENT ACTIONS.   Shortly thereafter, views in Shenandoah and the appraisals of them changed. By the late 1980s, the park had hired an outside firm to provide a computerized system for determining and monitoring decreased visibility and the sources of the pollution that could be causing that diminution. The public also noticed. A 1988 history of the park reported that "the most conspicuous environmental problem is the decrease in visibility that has occurred."[54] By the fall of 1989, the *Washington Post* warned visitors to "take a good look because the views are changing."[55] What had happened?

Because the park was created from privately owned land, Shenandoah is an island surrounded by development. Few other areas of public land, such as national forests, exist as buffers between the park and nearby communities. Those communities generate air pollution. The pollution alters air quality in the park in three major ways. First, tiny suspended pollutants, nearly 85 percent of which are sulfate particles, scatter light in the atmosphere, thereby reducing visibility. The natural average visibility from the park throughout the year is about eighty miles. Today's actual average is fifteen miles. The problem is particularly intense in warm, humid periods such as Virginia's late summer and early fall months. Second, sulfates combine with water in the atmosphere and fall as acid rain, which can kill vegetation and aquatic life. Shenandoah's precipitation today is ten to twenty times more acidic than normal rainwater. The pH level in park streams has dropped in the past few years to 5.5 and even 5.0 in 1992. Reproduction of aquatic life is impaired when pH levels fall below 6.0. Once bodies of water acidify, recovery is unlikely. Third, ozone is created when sunlight reacts with polluted air. Ozone is invisible, but it destroys vegetation by destroying substances on leaves, thereby making plants vulnerable to other pollutants and diseases. The ozone levels in Shenandoah also increased during the 1980s.

What are the sources of these pollutants? Several sources contribute, but one dominates. Air pollution can be caused by the open burning of trash and brush, but the relative impact of local trash burning is minimal. Automobiles, particularly those entering the park, contribute, but according to research carried out by the NPS Air Quality Division, all the vehicles in the park in 1991 contributed about seventy tons of nitrous oxides. Coal-burning power plants nearby and in the Ohio valley generate sulfates and nitrous oxides. One single power plant in Virginia generates 10,000 tons of nitrous oxides a year. The largest contributors, by far, to air pollution in Shenandoah are nearby power plants.

Why have the amounts of these pollutants increased so rapidly in recent years? Virginia alone has issued permits for fifteen new power plants since 1987 and is currently reviewing applications for four more. Together these plants will generate nearly 40,000 tons of sulfur dioxide and more than 72,000 tons of nitrogen oxides a year. Other areas west and upwind of the park have also increased power production and the corresponding contribution to air pollution in recent years. The results have been dramatic. In 1988, the park reported air quality at unhealthy levels two days in a row for the first time ever. In 1990, the park began posting advisories to visitors at the entrance station and visitor centers warning of diminished visibility. In the summer of 1991, visibility was less than ten miles 35 percent of the time and more than thirty miles only 16 percent of the time. Some studies now predict that the park's brook trout will be lost as a result of acidified water in the next ten years.

How have officials responded? Hodel's refusal to send the list of scenic vistas to the EPA in 1985 previewed later political intervention into agency behavior on this matter. NPS officials in 1990, led by the managers in Shenandoah, protested the granting of permits for new power plants in the state of Virginia. The NPS claimed their research showed that the new plants would have an "adverse impact" on the park. Field managers like Superintendent Bill Wade were initially backed by officials as high as NPS director Ridenour as well as environmental groups. These protestations were the first of this kind from any national park in the country. On the other side of the issue, the state's Virginia Power company had capacity for 15,000 megawatts in 1991 but planned to add 5,000 more by the end of the decade. The largest demand on record was 12,697 megawatts once in 1989, but Virginia Power argued that the state was growing. Wade and other NPS employees came under fire from local residents who still resented the park's past intrusion on private properties.

Later in 1990, central support for the field position wavered. In congressional hearings on Clean Air Act revisions, NPS director Ridenour adopted the administration's party line: "The administration has submitted a comprehensive clean air proposal which includes a number of provisions which would generally improve air quality throughout the Nation, including the air in the Nation's parks, obviously. This comprehensive, balanced program is perhaps the most effective tool in preserving air quality in the parks. . . . We believe significant progress can be made in decreasing haze without additional legislative authority."[56] In other words, the administration and its political appointees felt that specific concern for situations such as that facing Shenandoah was not necessary. The EPA backed the state of Virginia's assessment that, while air quality in Shenandoah had indeed worsened, the effect of the new utilities would be negligible. The EPA did so even while admitting that some parts of the studies were not technically credible and did not address the effects of pollution in bad-weather conditions. In an unfortunate choice of words, NPS spokesman George Berklacy admitted, "It appears the air has to be cleared between the EPA and the NPS."[57] In 1991, Virginia approved the new plants. Ironically, the state that had bequeathed the land to create the park had now sanctioned processes that were damaging it.

Shenandoah managers continued valiantly on their own. While they were still studying the effects of increased power production, park officials were stopped from testifying against pending permits for more plants in 1992. In April, according to Shenandoah management assistant Sandy Rives, Vice President Quayle's Council on Competitiveness convinced the secretary of the interior to stop the protests. I asked Rives if the secretary's action came in a letter or a phone call. "For the Super to testify, he needed approval from the Secretary," Rives responded and then continued with a comment that recalled Hodel's failure to send the vistas list to EPA seven years before: "Somehow it got lost."

CURRENT SITUATION.   Today the new permits are on hold. The EPA is now, according to Rives, backing Shenandoah's studies showing significant effects from new power plants. Superintendent Bill Wade and other park officials have been applauded in some accounts for their efforts to protect the park. But existing power plants are still operating without having to retrofit their equipment with machinery to reduce pollution. On the day I visited the park in October, Rives told me, "We're sure the air quality over Shenandoah is deteriorating." As the day warmed up and the visibility diminished, so was I.

SUMMARY.   The air quality in many national parks is seriously affected by external sources of pollution. The events involving Shenandoah's air quality have occurred only recently. The pattern of reaction, however, was established much earlier. Protestations by park officials were overridden by higher government officials and politicians, in spite of scientific evidence that there was a reason for concern about preserving natural areas. Field managers were thus unable to generate the support they needed to protect the park from dangerous external threats.

Why You Can See the Forest and the Trees in Prince Albert

Even in a national park administration building, late Friday afternoons are a tough time to find someone in their office. Yet there we sat, Prince

Albert chief warden Paul Galbraith and I, as the clock ticked along past 5:00 p.m. Paul is intense and articulate, and he was obviously caught up in the subject matter of our conversation. He did not even seem to hear the commotion outside as the other staffers began their weekend plans. He drew a circle on a piece of paper and then a smaller circle inside that as he talked. The circles represented a national park in a larger ecosystem. He drew little arrows from the smaller circle into the larger one. "What we have to do," he said emphatically, "is think our way into the ecosystem rather than thinking our way out of it. We're not an island." When I asked him how, I knew we would be there a lot longer. The attitudes of Galbraith, Superintendent Dave Dalman, and other staffers are transforming Prince Albert into a park that is more than just affected by external threats. Instead, the park changes the external threats.

BACKGROUND.   Prince Albert was established in 1927 in large part in gratitude by the recently elected prime minister for support from his local constituents. The park has been known since its creation as "one of the main gateways to the great hinterland" or as "the southern edge of the Great Northern Forest."[58] Prince Albert consists of nearly 4,000 square kilometers of lakes, grasslands, and boreal forests of coniferous and deciduous trees.

Prince Albert's mixed history does not necessarily encourage expectations that it might be an ecological pioneer. Until 1979, the park was managed as a recreational playground with emphasis on the town site of Waskesiu and its resort-like attractions of lakefront, hotels, and golf course. Adjustments in the water levels of the lakes are an example of how emphasis on visitor use threatened the preservation of natural resources in the park. In some areas, water levels were artificially raised to facilitate boat passage. These actions damaged lake-trout spawning areas, eroded shorelines, altered plant life on newly flooded areas, and increased siltation into lake waters. At the same time, however, the park became, largely through the efforts of noted naturalist Grey Owl, a wildlife refuge. As a result, today the park is home to caribou, deer, bear, elk, beaver, bison, and 227 recorded species of birds, including the famous white pelicans.

Between 1979 and the late 1980s, Prince Albert's managers changed policies, but only to a limited degree. As in so many parks in Canada, emphasis in Prince Albert shifted toward preservation. Park officials recognized overemphasis on development in early years and changed the orientation of the park and the town site. Managers, no longer focusing simply on recreation, used zoning, wildlife control, aquatic systems management, and redesign of the town site to shift the park to a more natural representation while still allowing it to supply essential services. For example, park managers ended the problematic water-level manipulations and began seeking appropriate ways to reverse the damage. Even tough critics like park historian Bill Waiser applauded the shift. Like so many parks in Canada and the United States, however, Prince Albert managers did not change their internal-external orientation. External threats, including development of the surrounding lake district, nearby logging operations, wildlife kills outside park boundaries, and global problems such as climate change, were largely ignored as managers looked inward at the park. Regional ecosystem management, for example, is barely mentioned in the 1987 management plan. This internal focus of Prince Albert and other parks has led some critics to conclude, "When all is said and done, a lot more is being said than done about ecosystem management."[59]

RECENT ACTIONS.   Management attitudes in Prince Albert have changed again in the past few years. The new attitude retains the preservation emphasis but expands the scope. This orientation involves a broader view of the park, not only a perception that a national park is part of a larger ecosystem and can be affected by it, but also a realization that the parks have a responsibility to educate and to encourage environmental citizenship. The two goals of ecosystem management and environmental education are key. Superintendent Dalman speculates that the change began about 1985 but admits that even now the park is in a period of transition.

How did this change come about? Both Dalman and Galbraith cite greater awareness of external threats as well as the empowerment of field employees through the recent decentralization of the CPS. "The strength of the organization is at the field and the front line," Galbraith said and then added, "And now *we* define our mission in terms of the health of the regional ecosystem and then think what can the park do about it." The structural changes that enable field definition of agency missions have occurred simultaneously with the increasing environmental orientation of employees. Interestingly, much of the philosophical basis for this new attitude in Prince Albert derives from the writing of an emeritus professor at the nearby University of Saskatchewan. When I was in Prince Albert, I saw an article by Dr. Stan Rowe in the park newsletter and quotations from his writing in the brochure on fire management. Galbraith gave me an article from a book by Rowe. One passage reads, "Wilderness, wild areas, can only survive as natural ecosystems that are ministered to, served, not just preserved, for only such attitudes on our part set their values above human needs and wants."[60] Prince Albert has responded to this call for action in several ways, particularly with specific research programs.

One program that operationalizes the new attitude is called the Saskatchewan Forest Habitat Project. This project is intended to protect regional forests extending through and beyond park boundaries. The project uses such diverse partners as the park staff, Weyerhauser Canada, and the Saskatchewan Wildlife Federation. Indicator species ranging from moose to woodpeckers are being studied to determine habitat requirements necessary to maintain biological diversity in the forests of and surrounding Prince Albert.

A second program, begun in 1991, is directed at regional fire management. Having adopted concepts of regeneration, prescribed burns, and allowing natural fires, the managers of Prince Albert are now working on a multiyear study of the effect of fire on the boreal forests across northern Saskatchewan. The study involves other provincial resource agencies and the public. The agency's goal is to use fire to help "restore and maintain a healthy and integrated ecosystem."[61]

The scope of a third project extends well beyond the region. The Boreal Ecosystem-Atmosphere Study (BOREAS) unites Prince Albert managers with officials from the U.S. National Aeronautics and Space Administration and with American and Canadian scientists. BOREAS comprises sixty to seventy experiments designed to improve understanding of the relationship between global atmospheric changes, such as the increase of greenhouse gases, and boreal forests. The experiments are concentrated in two locations, one being Prince Albert National Park. The project will use four research towers, designed to be minimally intrusive, to trace the exchanges of gases between the atmosphere and the forest ecosystem in the park. BOREAS is scheduled for six years (1990–96), with the bulk of the field work taking place in 1994. The government has been, according to Dalman, quite supportive.

CURRENT SITUATION.   All three described projects are under way. Both Dalman and Galbraith spoke with pride about existing efforts, particularly the park's participation in the BOREAS project. More outreach programs will be initiated in the near future. The park's management plan is being rewritten to focus all of park operations to reflect the new attitude. Galbraith is in charge of the rewrite. I asked him if the new, outward-looking orientation was a lot for the agency to handle. He responded, "The question is, What are we not going to do? Some services will probably be sacrificed, and we'll ask the public to do more." Then he added with a rare smile, "But there will be changes."

SUMMARY.   CPS management of Prince Albert has changed. During the past decade, park officials have shifted their emphasis from recreation to preservation. In the past few years, their efforts have involved the environment outside the park. That outward focus will make it easier for the park managers to use research not just to address, but possibly to help prevent, dangerous external threats such as global climate change in the future.

Even after leaving Prince Albert, I still wondered about both a practical matter and a theoretical question. Practically, how many Canadian parks are addressing external issues? Theoretically, why would decentralization assist field managers in dealing with external threats? Would they not be less effective than higher-level officials with wider jurisdiction? The programs at both Yoho and Kootenay answered the first question by showing that Prince Albert was not alone in ecosystem management. The statements of their managers answered the second. Both Yoho's Church and Kootenay's Jacobson meet with other regional officials or sit on area committees. Church said, "Before, we would sit on them [committees], and there was a philosophy that we needed industry because industry would bring us visitors, but now we're more worried about environmental issues than visitors." Still referring to regional meetings, Jacobson said, "Now when we sit down, they respect us." Again, environmental concern and decentralization have together changed the CPS.

Discussion

Managers in both Shenandoah and Prince Albert have attempted to use research to address external threats. Officials in both parks recognize and are motivated to deal with external issues. The efforts of Shenandoah's Wade, Rives, and other managers have been abandoned or even overruled by politicians and higher officials in their own organization. Meanwhile, whether endorsed, supported, or even left to their own devices, Prince Albert's Dalman and Galbraith have been able to shift the focus of their park and the agency in a new direction, outward.

Summary

The cases presented above obviously cannot depict all of the activities of either park service, but they do illustrate general patterns. As decision-making in the NPS has come under increasing scrutiny from politicians and higher administrative officials, field-level efforts have been increasingly overruled or countermanded. Grant Village, Fishing Bridge, motors in the Grand Canyon, traffic in Yosemite, and smog at Shenandoah are monuments to the politically induced shift of agency priorities away from preservation. As the CPS managers, particularly at the field level, have gained authority, however, many of the agency's efforts have shown a greater orientation toward preserving natural conditions. Elk Island's diverse wildlife, Jasper's backcountry, and Prince Albert's external orientation all reflect a combination of field-level initiative, scientific awareness, and environmental concern. The verdict is not yet in for the changes occurring in the town of Banff. These patterns may or may not reflect a more intelligent approach, but they do illustrate the effects of recent agency changes on field programs.

Chapter 5

With Politics in Mind

"The Park Service is now making more decisions with politics in
mind as opposed to what's best for the resource."
—*Rocky Mountain assistant superintendent Sheridan Steele*

When I stopped, I was deep in the Hoh Rain Forest of
Olympic National Park. I had been hiking along the Hall
of Mosses Trail on a near-perfect summer day when a shower of pine
needles cascaded down like green snow. The air was quiet and still on the
trail, but a breeze was blowing through the treetops hundreds of feet
overhead, gently shaking the needles from the ancient pines. Just enough
sunlight filtered down through the branches to warm the cool moisture
of the ferns and mosses on the forest floor. I had not planned to stop there,
but suddenly I could not move. An image popped into my mind of a black-
and-white photograph I had seen twenty years earlier in an old, yellowed
travel guide, a photograph of small children dwarfed by giant redwoods
in a Pacific Coast forest. I remembered thinking how peaceful and serene
the setting had seemed. For several minutes, I stood, transfixed by the
beauty of the forest and the feelings that such places inspire. Then I wondered, would it look the same twenty years from now?

Summary of Findings

This project began with some basic questions. Why has the National
Park Service (NPS) recently had so much difficulty preserving the national
parks, while the Canadian Parks Service (CPS) has become increasingly
focused on the very same goal? In broader terms, why would two similar
agencies interpret the same mission differently? Why does agency behavior change? I proposed a theoretical explanation and then presented the
evidence.

Theoretical Expectations

In chapter 1, I proposed a theoretical argument to explain changes in
agency behavior. The ability of a public agency to pursue a mandated goal

193

is determined largely by the political consensus behind the goal, the political support provided the agency, and the commitment of employees to the goal. When the levels of commitment from employees in two different agencies are similar, changes in behavior by the agencies are determined by changes in the political factors. If the levels of political consensus and support are increased, agency employees can structure their operations and increasingly concentrate their behavior on the pursuit of specific goals. Even difficult goals requiring long-term focus become attainable. If consensus and support decrease, agency operations become subject to second-guessing and external intervention. Power shifts to political authorities, and agency behavior increasingly reflects the short-term, electoral motivations of those politicians rather than the long-term concerns of the agency.

Empirical Analyses

The experiences of the NPS over the past fifteen years are consistent with the theoretical argument proposed in chapter 1. Changes that have occurred in the CPS support the argument, but with some important qualifications.

THE STATE OF THE NPS.   Chapter 2 described the heightened conflict over the preservation goal in the NPS mandate and the diminution of political support for the agency. These political factors led to changes in the NPS operating environment, including centralization of decisionmaking authority, diminished field-level morale, and intervention, to the point of micromanagement, by members of Congress and presidential appointees. The organization's behavior shifted to reflect short-term, electoral concerns. System expansion, agency funding, management policies, and the field-level programs described in chapter 4 exhibit the shift of emphasis from preservation of the natural resource to use of the parks for political purposes.

The evidence for both the structural and the behavioral shifts is overwhelming. Considerably more than 80 percent of agency employees interviewed, at all levels, say that decisionmaking is now more centralized and subject to political pressure than in the past. The concerns expressed in the quotation at the beginning of the chapter provide a sharp contrast to the opinions of the 84 percent of NPS employees who were optimistic about the agency's future at the end of the 1970s. Employees within the NPS itself, to their credit, have offered a brutally frank assessment of the agency's ability to pursue its mission for the twenty-first century: "Perceptions exist among many employees and observers—and not without bases in reality—that good job performance is impeded by lowered educational requirements and eroding professionalism; that initiative is thwarted by inadequately trained managers and politicized decision making; that the Park Service lacks the information and resource management/research capability it needs to be able to pursue and defend its mission and resources in Washington, D.C. . . ."[1]

External appraisals are also critical. The Conservation Foundation criticizes the centralized structure of the agency: "There is danger, however, that the superintendents' freedom to respond to the remarkably wide array of problems that arise throughout the diverse system could be overly curtailed."[2] The National Parks and Conservation Association (NPCA) says, "Overt and covert political intrusion threatens the very existence of our parks."[3] A wide range of observers generally agrees, as a review of nine important recent books says, "While the declared mission of the NPS has not changed, the de facto policies of the NPS have been responsive to politics, and so also have agency-developed natural resource and visitor management policies."[4] A representative appraisal concludes: "Over the course of years, however, a professional agency has been transformed into a political agency, leading to an emphasis on recreation."[5]

These shifts have neither pleased field employees nor used their skills. Expertise and knowledge gained only through close and continued contact with the parks have often been ignored, neglected, or overruled by political calculations. Many careerists, such as Shenandoah's Bill Wade, have continued to try to preserve what natural conditions remain and have occasionally succeeded. More often, their attempts have ended in frustration. Veteran employees, such as Lorraine Mintzmyer and Jerry Schober, have left or, like Howard Chapman, been forced out. Those who remain, such as Mike Finley and Sheridan Steele, openly express hope that the shifts of the past fifteen years can be reversed.

THE FUTURE OF THE NPS.   Recent political events offer some reasons to think that the agency may be about to undergo a soul-searching such as that seen in the CPS in the past decade. Agency documents, like *National Parks for the 21st Century*, express internal recognition of recent trends. Some politicians, like Bruce Vento, Democrat of Minnesota, continue to propose structural changes to achieve preservation ideals. In the past twelve years, these arguments have been heavily outgunned, but political calculations could change with the Clinton administration.

The Clinton administration began by appointing Bruce Babbitt, who was supported by environmental groups, as secretary of the interior and George Frampton, president of the Wilderness Society, as assistant secretary. Early in his tenure, Babbitt delivered a preservation-oriented speech to the NPCA in which he discussed such issues as systematic expansion, more emphasis on science, and ecosystem management. In an interview,

Babbitt agreed that the NPS had shifted too far toward development and away from preservation. He took some specific action by stopping the construction work at the Manassas stables described in chapter 2. President Clinton requested and received an increase in appropriations for the NPS for fiscal year 1994. Some preservation advocates are enthusiastically hopeful that Babbitt will lead a fundamental reversal in national park management.

Reversal of years of political intervention into the agency is hardly imminent, however. Political influence on the NPS has come from both parties. Examples are the influence of senators Alan Simpson and Malcolm Wallop on fire policies in Yellowstone and of Senator Robert Byrd on appropriations. The Clinton administration has shown a willingness to compromise on public lands issues, retreating in the face of western congressional opposition on promises to raise grazing and mining fees. Babbitt's own history in Arizona shows a readiness to deal with developers, industrialists, and other interests that may challenge a preservation emphasis in the parks. In addition, the new head of the NPS, Roger Kennedy, is an experienced bureaucrat (formerly with the Smithsonian) and a television journalist (with NBC), but not a park service careerist. In fact, the choice of Kennedy over Paul Pritchard, president of the NPCA, signals a more moderate position in itself.

Even if presidential demands change with a Democratic administration, changing NPS behavior will also depend on the actions of other relevant participants, particularly members of Congress. Recent studies of other public agencies suggest that congressional micromanagement has occurred even in the face of aggressive executive behavior. If the Clinton administration attempts to change the NPS without congressional support, the agency will remain vulnerable to intervention. Even if that support can be generated, persuading park careerists that legislators are genuinely concerned about preservation issues will be, as congressional sources admit, difficult. The agency will have to contribute to changes by educating and training workers in such matters as congressional liaisons. These are the political realities that must be overcome if the management of American national parks is to be changed.

THE STATE OF THE CPS.   As chapter 3 described, the political environment of the CPS has also changed dramatically during the past fifteen years. Though political authorities were not consistently supportive with resources or commitment, they verbally clarified the CPS goal of preserving natural resources. The CPS reaped benefits of insularity and relative autonomy from being transferred to a supportive cabinet department and from the increased deference to bureaucratic expertise that occurred during the 1980s. Given a rhetorical green light and increased political support,

CPS careerists shifted many responsibilities to the field level, improved agency morale, and used their growing discretionary authority. As a result, the CPS has reversed decades of orientation to recreation and commercialism to develop a scientific system of expansion, management policies with a focus on preservation, and innovative programs such as those described in chapter 4. The changes have been particularly acute in the Western Region and increasingly noticeable during the past few years.

Again, the evidence is compelling. Virtually 100 percent of the CPS employees I interviewed, at all levels of the organization, say that decision-making is now more decentralized and that the agency has turned the corner to pursue a more preservationist focus. The department's own internal appraisal of parks in the twenty-first century says: "National parks and other protected places are our lifeline to an ecologically stable future. They are places where the forces that animate our planet and make it unique are allowed to operate with minimal interference by man. . . . They are places where people can study the vital functions of nature, to understand better the potentially threatening human impacts elsewhere. They do not impede our future. They anchor it. They are not a repudiation of economic development. They balance it."[6]

External appraisals are mixed. CPS watchdog groups remain wary after decades of exploitation but express guarded optimism about recent changes. Some advocates perceptively realize that the changes in the agency are taking place faster at lower levels of the organization and encourage system-wide acceleration to catch up. More impartial observers applaud various programs of the CPS and cite shifts in agency orientation away from commercial exploitation and toward conservation of natural resources for their own sake. Recent scientific endeavors have applauded CPS assistance in use of information available only at the field level. Evidence for the CPS shift is not yet widely acknowledged because changes are under way.

The qualifications to support of the hypothesis are important. The statement that an agency has moved in one direction does not mean that all outcomes reflect that shift. The CPS is still burdened by a legacy of emphasis on recreation and development. The continued existence of towns in parks, the slow removal of logging from Wood Buffalo, the allocation of budget dollars primarily on the basis of visitor totals, and the unlikely completion of system expansion by 2000 are all reminders that the agency faces difficult challenges in reversing past orientations.

THE FUTURE OF THE CPS.   The future actions of the agency will also be affected by electoral events, including the public's reaction to various attempts at constitutional reform and the return to power of the Liberal party in 1993. Even where changes have been initiated and accepted in the field, implementation and consistency still depend on support from higher levels of government. With constitutional disarray, that support may be difficult to secure in the future. The increased strength of provincial governments, with their own concerns for economic growth, enhances the need for park managers to continue nurturing external relations and developing an outward focus.

Nevertheless, the autonomy of the CPS and the corresponding delegation of authority to park managers is only likely to increase in the coming years. Current fiscal realities and general emphases on decentralization will ensure continuing empowerment of trail-level bureaucrats, no matter how the elections and the fight over constitutional unity are decided. Further, CPS employees are committed to the ongoing evolution. As Richard Leonard of the Prairie Regional Office said, "The top-down approach is a dinosaur." The actions taken at lower levels of this decentralized organization will continue to determine the future of the Canadian park system.

COMPARISON.   Comparison of these two agencies with similar missions, threats, and employee commitment shows how they have recently changed and their behaviors diverged. CPS managers have not only moved in front of the symbolic bulldozer mentioned in chapter 2, but they now often stop it. NPS managers are forced to follow in its tracks, maintaining the new developments mandated by others for the parks. In the words of those who know best, the managers in the field, the NPS is now forced to manage parks "with politics in mind," whereas the CPS is showing, in many cases, how to "be smart" in the pursuit of preservation.

Broader Implications

This study has implications for the national parks and beyond. The empirical results bear on important theoretical questions within this context. The theoretical arguments can be extended to other public agencies within Canada and the United States.

Theoretical Questions within the Parks Context

The management of North American national parks illustrates important theoretical issues at each of the three stages of analysis used in chapters 2 and 3. Those stages are the political environment of national park agencies, the operating structures, and the policy results.

CHANGES IN THE POLITICAL ENVIRONMENT.   Figure 5-1 displays the factors in the political environment affecting recent park service developments

FIGURE 5-1. *Political Determinants of Agency Pursuit of Goal*

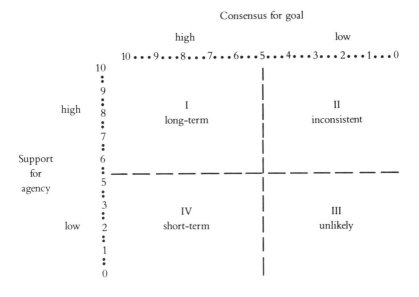

in a two-dimensional scalar diagram. The important variables of goal consensus and political support are shown as continuous variables ranging from high to low. The Roman numerals represent cells, which are separated for purposes of explanation by arbitrary dashed lines. The preceding analysis supports the hypothesized outcomes depicted in cells I and III. When political consensus and support are high (cell I), the agencies will be capable of consistent, long-term pursuit of a specific goal like preservation. When consensus and support are low (cell III), pursuit of the preservation goal will be difficult. Agencies characterized by cell II pursue specific goals only inconsistently, depending on powerful interests that fill the vacuum left by the lack of consensus. Agencies described by cell IV may be effective in pursuing a specific goal, but their actions are only short-term, determined by dominant political interests on which the agency depends for support. Several questions arise from this theoretical framework.

First, are these outcomes consistent with the theoretical literature? While no single work has characterized the political environment in quite these terms before, a variety of pieces suggest that these outcomes should not be surprising. Social scientists from Theodore Lowi to James Q. Wilson have posited the importance of legislative mandates. As chapter 1 discussed, agencies are more likely to pursue goals that are clear, consensual, and explicit. Other analysts have emphasized the importance of institutional structure. These writers suggest that relative autonomy and supportive agency placement can facilitate an agency's pursuit of its mandated goals. When these factors change in ways that reinforce each other, as they have in the cases of the NPS and the CPS, the resulting changes in agency behavior are dramatic.

A second question that naturally follows concerns whether or not these developments produce an equilibrium. In other words, are the outcomes depicted in figure 5-1 static or dynamic? I would argue that the behavioral patterns described are not static, but are likely to increase and intensify.

In the case of the NPS, as the agency strays further from its preservation mandate, its actions attract more attention, increased criticism leads to greater political involvement, and the agency becomes increasingly responsive to political concerns. Debate over agency goals becomes increasingly conflictive, and support diminishes. When I asked NPS employee Steele, "What if trends continue," his exact response was, "Some day all decisions will be made on the politics of the moment, not the agency's mission." He was speaking for many who fear the upward spiral affecting the agency in spite of potential changes that could result from the Clinton administration or from other political developments. Furthermore, employee commitment to long-term goals may wane as morale diminishes.

An agency's pursuit of mandated goals solidifies its focus, and effective pursuit strengthens political support from other external participants. Employee morale and commitment soar. Again using an example from the field, Prince Albert superintendent Dave Dalman told me that the CPS will continue to move in the direction of emphasizing preservation if for no other reason than personnel. Employees whose early careers were spent in the pre-change, pro-use predecessors of the CPS are being replaced by resource-conscious individuals hired by people like Dalman. Reinforcement of changing behavior by replacement of personnel with those sympathetic to the continuing changes is a phenomenon that has been noted in other public agencies. Relatively autonomous agencies have more freedom to choose personnel sympathetic to agreed-upon goals.

The next obvious question is, therefore, how to reverse these trends. Suppose that reformers wished to return the NPS focus to the preservation goal. Could a Clinton administration, or other relevant participants, reverse the effect of the political environment by either building a consensus on goals or increasing institutional support? Certainly, those efforts would be received well and assisted by the majority of preservation-minded employees of the NPS. Neither alone would be sufficient, however. If policymakers emphasized preservation while leaving political support unchanged (cell IV), agency efforts would still suffer from underfunding, frequent external intervention, and interagency conflict. If political support were increased but goal emphasis remained conflictive (cell II), NPS actions would continue to be second-guessed and contradicted by relevant authorities. For example, making the NPS an independent agency, as many have called for, without focusing agency behavior on preservation will not produce a dramatic reversal of current trends. In general, changing agency behavior requires addressing both political factors.

EFFECTS OF THE OPERATING ENVIRONMENT. The effects of the operating environments on the actions of these agencies demonstrate four important theoretical issues: the interdependence between appointees and careerists, the role of internal structure in efficacious policymaking, the relationship between internal structure and political sensitivity, and the stimulation of political interest in agency behavior that follows potential revenue increases.

The recent dialogue about the relationship between appointees and career civil servants has evolved into a debate over the costs and benefits of increased political control. The recent history of the NPS illustrates several facts that are central to this debate. First, increased political control will alter behavior. Agencies that are more responsive to appointees (such as the NPS) are likely to display the short-term, electorally motivated behavior that can satisfy demands of political authorities. Second, giving appointees greater control over agency activities (as in the NPS) will likely heighten dissatisfaction among careerists and may lead to a corresponding drop in morale. Third, increased presidential assertion through political appointees does not preclude, and in fact may inspire, congressional attempts to micromanage agency behavior. This latter outcome is particularly likely in agencies, such as the NPS, that have been fragmented by inner divisions and are thus vulnerable to access and control by members of Congress. Those who advocate strengthening presidential control over agency behavior must weigh these potential costs against the benefits of pursuing responsive administration.

A second issue concerns the effect of internal agency structure on the efficacious use of knowledge by organizations. Many noted scholars have long argued for increased use of decentralized authority, under certain circumstances, in organizations to make better use of local knowledge and to apply general principles to specific situations. Particularly when agencies pursue agreed-upon goals in diverse situations, decentralization can be an intelligent response.[7] Both park service agencies use some decentralized authority, but the CPS has shifted heavily in that direction in recent years. By empowering committed field-level employees, public agencies like the CPS can stimulate motivation, dedication, and innovation in specific situations that defy system-level responses. The high-tech scientific endeavors developed at Elk Island and Prince Albert are examples. Limitations to decentralized authority are obvious, however. Field efforts bereft of central support and resources are doomed to failure. The slow pace with which the Canadian government pursues final system expansion illustrates the limits of reliance on decentralized initiative.

A third and related theoretical issue recalls one of the oldest lessons of the public administration literature. Agencies are never insulated from politics, whether centralized or decentralized. The effect of political considerations on the centralized NPS was illustrated time and again, but the lower levels of the decentralized CPS also face political realities. Certainly, the potential for excessive local political influence on low-level employees of decentralized agencies is legendary. The CPS experience at Banff, first unable to control the town site and then turning over power to the town, recalls the argument that decentralized agencies must be wary of powerful local constituencies.

Nevertheless, some organization theorists have argued that, counter to certain capture-related expectations that local influence over decentralized agencies will only preserve the status quo, delegating politically sensitive decisions to field employees and away from politicians can be beneficial in stimulating agency responsiveness to specific situations. The experiences of several CPS managers, particularly the relations between them and town councils in places like Jasper and Prince Albert, support the point. The argument is strengthened by a fairly surprising finding in negotiations over challenges arising outside of a field manager's jurisdiction. While one might expect that decentralization would diminish a park manager's ability to call on higher authorities, in fact delegated power increases the respect that he or she receives from the outside challengers. Politics are not removed from decentralized organizations. Thus field employees, if they are to pursue the statutory goals of the agency, need the resources and the authority to address political challenges.

Finally, a fourth issue involves the potential increase of resources provided to an agency that can follow an increase in user fees or service charges. As chapter 2 mentioned, the NPS of the 1980s joins a line of public lands agencies that have experienced greater political circumspection following increases in the charges they levied on users. When they raise fees or charges, public agencies are not guaranteed higher revenues from the money that is funneled back through the general spending process. Rather, they risk attracting greater interest from legislators who see the potential to have their name attached to some of the funds that do return to the agency. Thus a public agency may end up with the same amount of funding but with more strings attached after a fee increase.

POLICY OUTCOMES.   The divergent policy behaviors displayed by the two park services have implications for several broad issues related to policy outcomes. These issues include differences between Canada and the United States, the nature of policy changes, and methods of policy evaluation.

First, the different behaviors of the park services should not be interpreted as affirmation that the Canadian system is simply more effective than the American in matching policy results to policy intentions. Though one could find counter examples in other policy areas, contrary evidence is available even within the studies of the national parks. Before the changes of the past fifteen years, the actions of the NPS were more consistent with its original mandate than were those of the forerunners to the CPS. The recent shifts of the NPS away from and the CPS closer to statutory intentions of preservation were caused by the changes in goal consensus and political support described earlier. These changes were certainly affected by societal developments, but the outcomes are not permanently guaranteed by institutional differences.

Second, recent policy developments in the park services are logical outcomes of changes in important variables. Contrary to what some models suggest, organizational choices and decisions were not random results. The recent developments involving the park services are also not simply the incremental adjustments predicted by some compelling theories. Rather, recent changes have been and continue to be fairly dramatic. Significant shifts in political variables can lead to important policy manifestations, even reversals of agency emphasis as in the case of the CPS.

Third, the analysis offers a straightforward way to evaluate an agency's behavior by offering an argument for how well an agency will be able to pursue a specific goal. To reiterate, an agency's ability to pursue a mandated task depends on political consensus, political support, and the commitment of employees to that task. This formulation enables the causes of unsatisfactory agency behavior to be traced to these variables.

The Analysis Related to Other Public Agencies

How far can the analysis be extended beyond the national park services to other public agencies? Thorough consideration is obviously beyond the scope of this work, but I suggest that the theoretical framework depicted in figure 5-1 can be readily applied to other public agencies in Canada and the United States. These agencies all experience varying amounts of consensus on specific goals, different levels of political support, and perhaps even contrasting degrees of employee commitment. I argue that effective pursuit of specific agency goals will depend on high levels of all three.

I offer some brief illustrations by applying the framework to several U.S. public lands agencies. These organizations have had to deal with difficult challenges similar to those facing the CPS and NPS in lands and waters owned by the general public. I use three brief examples (one in cell II, one in cell IV, and one with shifting employee commitment) from different agencies in different time periods to suggest how the argument can be generalized. These accounts are just snapshots of particular agencies at specific times; more thorough analyses could describe the shifts of agencies from one cell type to another.

THE GRAZING SERVICE.   This brief history of the U.S. Grazing Service (GS) illustrates the application of the theoretical framework to an agency's behavior displaying cell II characteristics. The GS was created in 1934 by the Taylor Grazing Act to manage the public range land for intelligent, sustainable use. The act authorized the agency to do whatever was necessary to carry out its mandate. The lack of political consensus on the meaning of sustainable use was apparent in that the definition depended on who was asked. On the one hand, western ranchers saw an agency that could inexpensively help them avoid overusing range resources. Indeed, the agency was mandated to "cooperate" with local stockmen's associations. As other sources have noted, western ranchers would have given up those benefits rather than agree to any drastic changes in grazing rights, regulation, or higher fees. On the other hand, eastern congressional representatives envisioned a revenue-raising agency generating windfalls from higher grazing fees. To this end, the agency was entitled to set precise fees and regulate the numbers of animals using the range.

In spite of the lack of consensus over whether the agency was to regulate or cooperate with stockmen, the GS was well supported in its early years by political actors. The agency enjoyed a period of little political interference and steadily increasing appropriations between 1935 and 1944. GS appropriations until 1943 are depicted in the middle graph of figure 2-2. In this period, the GS exemplified an agency in cell II of figure 5-1, with nonconsensual goals but high political support.

Perhaps consistent with the expectations of the capture literature, the void in consensus for agency goals was quickly filled by a nurturing and cooperative relationship with powerful economic constituents, stockgrowers and ranchers. Consider the fees charged by the agency in the period between 1936 and 1946. While Forest Service fees for comparable grazing rights grew from $0.13 per cow to $0.27 per cow, GS fees remained at $0.05 per cow. As one member of Congress remarked, "it is common knowledge that they [cowmen] practically have been running the Grazing Service."[8] Not surprisingly, GS establishment of intelligent grazing procedures, the goal over which consensus had not been reached, was inconsistent in this period at best, often ignoring studies recommending sizable grazing reductions. As others have noted, "The resolution of the public domain question proved to be an intractable one," leading to intense political controversies.[9]

The mid-1940s witnessed a slide of the GS from cell II to cell III. When the agency did propose raising grazing fees to reduce overuse, the GS came under fire from ranchers and from western Senators. At the same time, however, eastern representatives intensified their criticisms of low fees and charges that the agency had been captured. By 1946, political support for the GS was gone. The agency was caught between "a Senate committee which refused to allow increased fees [and] a House committee which cut appropriations because fees were not raised."[10] Unable to pursue an unclear mandate or to satisfy diverse political interests, the GS was merged with the General Land Office in 1946 to become the Bureau of Land Management (BLM). Diminution of political support is evident in both budget and personnel for range land management, which were slashed by more than 60 percent. The BLM has since been generally regarded as politically weak, underfunded, and often ineffective.

THE CORPS OF ENGINEERS.   Since the early nineteenth century, the mandate of the U.S. Army Corps of Engineers has been to assist the nation's economic growth by managing the nation's navigable inland waterways. I argue that for much of the twentieth century, the agency's pursuit of this goal displayed cell IV type characteristics.

Political consensus was high for simple, electoral reasons. Much of the Corps's activity involved the construction of dams and levees to channel waterways. Members of Congress could use these public works projects to pump money into their local economies, to create jobs, and to provide visible benefits to constituents. The Corps allowed these projects to be spread geographically. This was possible mainly because of logrolling, a practice in which each member would support (or at least not oppose) projects in other members' districts. Congress also mandated cost-sharing arrangements in which the federal government paid up to 100 percent of construction costs. As a result, the Corps received sizable budgets to carry out projects.

Because the political-support variable includes agency autonomy, however, I classify it as low. While congressional involvement enabled the Corps to develop a certain independence from executive branch directives, it also enabled intense political intervention into agency decisions. As the Corps became the engineering arm of Congress, individual members increasingly directed, specified, and funded Corps behavior. This case illustrates that budgets alone are not enough to ensure or deny political support to an agency. The Corps was not autonomous from Congress but was, rather, part of a symbiotic relationship involving the production of questionable public works projects in return for sizable appropriations.

Congressional intervention fostered an impression that the Corps pursued projects that were useful to individual legislators but were also often economically inefficient and frequently detrimental to environmental quality. Corps behavior responded to short-term, political demands rather than to long-term concerns about economic growth using the nation's waterways. The Corps was often a willing participant in these matters, but to be fair, it did not always call the shots. For example, the recent floods in the Missouri River valley were anticipated by the Corps in the 1940s. Corps engineers advised Congress that protecting floodways along the river from development would be a wiser, long-term solution than building dams and levees. They were, however, outlobbied by agricultural, real estate, and railroad interests that persuaded Congress to omit flood-ways in their flood-control plans. These areas have since been flooded several times, causing considerable loss of life as well as economic damage.

This reputation stimulated well-documented recent efforts to reduce congressional manipulation by shifting responsibilities to professional staff relying on input from the public. It is not yet clear whether these efforts to make the agency more professionally responsive and less responsive to congressional pressure have been successful. Many see the agency as still serving its real master by pursuing the short-term political desires of members of Congress.

THE FOREST SERVICE.   The U.S. Forest Service (FS) in the 1980s provides an ironic reminder of the importance of employee commitment to agency goals. In the cases of the Grazing Service and the Corps of Engineers, employee commitment was assumed to be constant largely because deter-mination of variations is historically difficult. Furthermore, political var-iables alone significantly affected the behavior of the GS and of the Corps. The FS has recently experienced a change in the factor once presumed to be the agency's greatest strength, the commitment of employees to the multiple-use goal.

Traditionally, the FS displayed cell I characteristics. The FS actually developed its own primary goal before it was statutorily recognized. Since the influence of Gifford Pinchot, the first head of the agency, the FS has always preached a philosophy of using the forests for the greatest good for the greatest number. In practice, this philosophy was manifest as the mul-tiple-use doctrine: forests were to provide for timber, watershed mainte-nance, wildlife, mining, grazing, and recreation. This doctrine was offi-cially recognized as the agency mission in the Multiple Use-Sustained Yield Act of 1960. The FS also received high levels of political support, in budgets and in autonomy, based on the influence of well-connected timber constituencies and on the agency's reputation as a highly professional or-ganization. Even into the early 1980s, politicians rarely intervened in agency affairs. The FS developed into an agency in which employees at all levels consistently pursued the agency mission.[11]

In the past thirty years, however, and particularly recently, changes have affected this agency. In part, these changes derive from the Reagan administration's efforts to emphasize some parts of the multiple-use mandate more than others. In particular, the Reagan administration attempted to cut resources for recreation while often increasing production-oriented programs. Recent changes also derive, however, from the shifting views of agency employees. A substantial number of FS employees have grown increasingly critical of the agency's multiple-use focus, arguing that such a focus favors development of the national forests at the expense of ecological concerns such as endangered species and watershed management. The recently formed Association of Forest Service Employees for Environmental Ethics now claims more than 1,500 of the agency's 38,000 employees. This dissension in the ranks is quite a change from Kaufman's account of rigid "voluntary conformity" among employees.[12]

With these changes, the Forest Service's characteristics appear to be slipping from cell I to Cell III. The revolt within has shaken the political consensus on multiple use. The agency has been subject to a "tremendous increase in the amount and intensity of public involvement in Forest Service management activities."[13] The FS has been pulled back and forth by growing environmental demands and intensified timber industry pressure. Political support has diminished because politicians are now much more likely to intervene in agency affairs, as exemplified by recent investigations of the firing of activist John Mumma. As the agency's placement in figure 5-1 has shifted, FS emphasis, by most accounts, has become more oriented toward one specific aspect of its mandate, timber development. Other aspects of the multiple-use doctrine have been neglected as a result.

Recommendations

As I said in chapter 1, I consider preservation an important and worthwhile goal of the national park services. The national parks of Canada and the United States are special places. Few places on this continent have been set aside in a relatively natural condition. To have accomplished this with forests, rivers, and mountains that inspire our sense of wonder took superlative foresight. How can the pursuit of preservation be made more possible and more effective? I consider some proposals for changing the form of the park services and then make some specific recommendations for changing their behavior.

Proposals to Change the Institutions Governing the Parks

Because the parks are important to so many people, many observers have expressed concern about their current condition, offering recommendations or proposals. Several proposals consider changing the nature of the agencies themselves, replacing, restructuring, and reorganizing the current governing institutions of the parks.

REPLACING POLITICAL INSTITUTIONS.   In recent years, a number of proposals, often from economists, have encouraged either placing the parks in private hands or, at least, relying more on market forces in their management. Based on the premise that parks are often subjected to conflicting goals pushed by competing interests, these proposals suggest that privatization would simplify the management of competing claims. Economists argue that democratic institutions are often inefficient allocators of resources and incorrect assessors of true market prices for goods such as parks. Specific recommendations range from turning over individual parks to corporations such as Disney to retaining public ownership but maximizing the possessory interest of concessioners. I argue, using insights from several noted economists, that such intensive reliance on the market for management of parks would be problematic.

First, how can the market set an appropriate price for the preservation goal? Preserving parks in relatively natural condition provides benefits that are both difficult to measure and indirect. The aesthetic quality of a scenic wonder, for example, defies quantification, even through such tools as willingness-to-pay criteria or individual utility curves. Calculating the willingness of tourists to pay for visits to parks ignores the indirect benefits to those of us who derive some satisfaction from knowing the lands are there even when we are not visiting and from the future generations who will enjoy these places as well. How can private enterprise value these potential sources of demand? As economist Robert Nelson says, "No entrepreneur could readily convert benefits of this kind into profits."[14] Park service employees appreciate their pay in sunsets precisely because sunsets do not have price tags attached.

Second, establishing private ownership or control over national parklands is both impractical and undesirable because of the principle of nonexcludability. Parks illustrate one classic property of public goods in that being available to one, they are available to all. This nonexcludability makes private ownership impractical. As economist Robert Dorfman says, "There isn't much point in owning anything from which other people cannot be excluded."[15] Establishing private ability to exclude certain people from a park is also socially undesirable because it creates the potential for discrimination and inequity.

Third, national parks are unlike other market goods because they have no substitutes. While some might argue that other recreational resources provide proxies, two factors preclude substitutes for the parks. First, as noted earlier, each park unit is unique. Second, once the parks have been changed or drastically altered, they are gone forever. One need look no farther than the reservoir waters of Lake Powell or behind the O'Shaughnessy Dam to realize the permanent loss of such remarkable and unique places as Glen Canyon and the Hetch Hetchy valley. In describing sustainability as possible through the use of substitutes, even Nobel laureate Robert Solow admits, "That does not preclude preserving specific resources, if they have an independent value and no good substitutes."[16] The increasing crowds at such places as Banff and Yosemite are testimony to the lack of substitutes for these natural wonders.

Fourth, privatization of national parks would foster the potential abuse inherent in a "tragedy of the commons" situation. In his seminal article, Garrett Hardin describes how private ownership of a common land resource leads inevitably to ruin as each rational individual attempts to maximize his or her own personal gain, collectively destroying the common property.[17] Consider the possibilities for privatizing a national park. The park could have one dominant private owner or manager, an extension of the current NPS arrangement for single concessioners. Alternatively the park could have multiple operators, an extension of the current Canadian leasing system for park residents. The former situation would create monopolistic powers with all the inherent inefficiencies and abuses. The latter situation would create incentives for each owner or operator to attempt to maximize his or her own profits (for example, by putting up a larger neon sign) even while contributing to the demise of the natural conditions in the park (for example, turning Yosemite Valley into a Las Vegas strip). The latter situation is not that dissimilar from the overgrazing on public lands that has beset BLM management. The tragedies that can result from private overuse of commons lands have been recognized for decades. As Arthur Pigou argued, "It is the clear duty of Government, which is the trustee for unborn generations as well as for its parent citizens, to watch over, and if need be, by legislative enactment, to defend, the exhaustible natural resources of the country from rash and reckless spoliation."[18]

In summary, I do not dismiss the potential benefits of greater use of market forces in the parks. In fact, in specific recommendations below, I advocate certain changes, such as increased accountability, that emanate from economic principles. However, privatization of parklands is not the way to reform park management. First, the economic difficulties described above convinced even the godfather of capitalism that privatization of parks is a bad idea. Adam Smith wrote, "Lands for the purpose of pleasure and magnificence—parks, gardens, public walks, etc., possessions which are everywhere considered as causes of expense, not as sources of revenue— seem to be the only lands which, in a great and civilised monarchy, ought to belong to the crown."[19]

Second, it will not happen, if for no other reason than that politicians do not want to turn the parks over to private industries. Why would they?

Recall the comment by Superintendent Gaby Fortin in chapter 3: Parks provide opportunities for the good news. Indeed, the various seminars and groups advocating privatization in the early 1980s have been unable to generate the momentum and support necessary to push such dramatic changes.

RESTRUCTURING POLITICAL INSTITUTIONS. Some have considered restructuring the allocation of responsibilities for public land management between different levels of government. Such a proposal could conceivably include delegating the management of parks and historic sites to state and provincial governments. These governments already derive some revenue, although not usually in the form of taxes, from federal lands under their jurisdiction. How the lands are used is certainly important to regional economic activity, particularly tourism. Thus state and provincial authorities would have interests as well as local expertise in managing these lands. Furthermore, decentralization of various federal responsibilities is a politically popular ideal. Such a proposal in the area of public lands has received significant grassroots support, particularly from some western provincial authorities in Canada and from the sagebrush rebels in the United States.

Such proposals have rarely gotten past the consideration phase, however. First, parks and historic sites are not strictly regional in nature, attracting visitors from throughout both countries. Most U.S. and Canadian citizens favor systems of national parks that belong to everyone. Second, increased international visitation in recent years reflects the degree to which these sites have become national treasures for both countries. Third, one has to wonder how interstate or interprovincial competition might affect the parks. Would competition over tourist dollars sacrifice the preservation goal? Certainly the behavior of state governments in other policy areas suggests that such a dynamic can have important effects. In general, too many fear that transfer of federal lands to subnational authorities would result in lost expertise, economies of scale, and broad social purpose.[20]

While the notion of turning over national parks to state and provincial governments may not be practicable, however, some aspirations of the proposals can be retained. I refer below to the benefits of decentralizing management responsibilities.

REORGANIZING POLITICAL INSTITUTIONS. Several current proposals for reforming the park services involve reorganizing the managing agencies. Most of these have centered on the NPS. In some ways, these proposals date to the compromises made in the original creation of the agency. Many advocate moving the NPS out of the Department of the Interior and making it an independent agency. These efforts received partial approval from the House of Representatives in 1989. H.R. 1484, sponsored by National Parks Subcommittee chair Vento would have appointed the NPS director for a five-year term with complete control of the agency. Vento argued, "We have our parks being run by political flies . . . who make day-to-day decisions that are counter to the purpose for which we established these parks."[21] The Senate did not act on the legislation.

Other proposals call for citizen oversight boards. Vento introduced another bill recently to create a three-member national park system review board to "maintain a continuing review of programs and activities of the NPS."[22] If this proposal sounds similar to the 1916 suggestion of Steve Mather mentioned in chapter 2, that heritage did not overcome opposition from Republicans and the Department of the Interior. Another recommendation calls for using endowment boards, similar to those overseeing museums, composed of persons appointed by the president and confirmed by Congress who would manage parks as "trusts or endowments" for the public. These boards would enable discretion while, arguably, increasing accountability.

Finally, the wise-use movement takes a different track, advocating breaking down the NPS into smaller, more accountable units with narrower responsibilities. Its members argue that the current NPS is "a bureaucracy so huge and powerful that it can ignore the public will, the intent of Congress, and direct orders from the Secretary of the Interior with impunity."[23]

These proposals vary considerably in merit. The wise-use recommendation is based on a false premise, the empirical record showing that the NPS is anything but a powerful, runaway agency. Making the agency independent and appointing review boards are both worthwhile suggestions intended to provide the agency with increased political autonomy. Independence is no guarantee of such a result, however, and the agency would still face interdepartmental conflicts, particularly over the preservation goal, with agencies managing adjacent lands. Boards would add another level of political appointees who might or might not provide insularity for the agency. Indeed, personnel within the NPS have mixed opinions as to the ultimate effect of such proposals.

Proposals to Change the Behavior of the Agencies

My own recommendations concentrate on the behavior of the park service agencies. In this regard, these suggestions are more in line with recent proposals to revitalize or re-energize public agencies rather than to replace them with markets, state agencies, or endowment boards. Still, certain ideals evident in the proposals described earlier are present: accountability, decentralization, discretion, and autonomy. These are the qualities many in the CPS have been striving for in recent years. As many in the CPS admit, however, even their own efforts can be improved. The NPS needs even more dramatic change.

In making specific recommendations, I retain the focus on the preservation goal. I reiterate that by arguing for preservation, I do not advocate locking people out of the parks. When NPS director Roger Kennedy was asked about entrance quotas two days after taking office, he answered that keeping people out of the parks was not his goal and restated the theme that these lands, after all, belong to the public.[24] I concur. I do advocate use of the parks, however, only in ways that enhance the natural experience. Preserving the opportunity to enjoy nature on its own terms should be the primary explicit goal of these agencies.

In organizing these recommendations, I use the model of agency ability to pursue such a goal described in the previous chapters. Specifically, I offer recommendations to enhance the agencies' ability to pursue the preservation goal by increasing political consensus behind, political support for, and employee commitment to that goal.

BUILDING POLITICAL CONSENSUS.   If current trends affecting the NPS are to be reversed and recent changes in the CPS are to be institutionalized, political consensus must be built for the goal of preservation. Several steps would assist in that effort.

—Pass meaningful legislation making preservation the dominant and explicit goal of the park services. Analysts of government performance ranging from the pragmatic Grace Commission to academic theorists have recognized that an explicit sense of purpose is necessary for efficacious bureaucratic behavior. This is especially true in policy areas like public lands management, where so many interests express so many conflicting demands. Making preservation the explicit goal of the park services will be true to the original agency mandates and will also benefit from the fact that employees are already committed to this ideal. As Durant's study of the BLM shows, employee receptivity to explicit mandates is an important condition for political-bureaucratic cooperation.[25]

For the NPS, Congress must reiterate the original agency goals of using the parks only in ways that do not impair their natural conditions, as the Canadian Parliament has done in the Green Plan. Representative Vento has called on the legislature to do precisely that in his proposed National Parks and Landmarks Conservation Act. His efforts warrant support.

For the CPS, themes expressed in PS 2000 and the Green Plan need reiteration and reinforcement. For example, the new government will gain considerable credibility by showing meaningful progress toward the explicit timetable for expansion of the system. The system plan will only be successful if commitment is consistent even through changes in administrations and constitutional crises.

—Reorganize public lands to make the preservation effort more focused. A consensus will be more attainable if the responsibilities of each natural resource agency are clarified. This proposal is particularly directed at the public lands situation in the United States. The 90 million acres of wilderness areas (lands to be preserved in current condition) in the United States are currently managed by four different agencies. All should be under the jurisdiction of an NPS with renewed focus on preservation. Indeed, Vento's proposal for a wilderness division chief within the NPS would facilitate such a transfer. Recreation areas, currently split between the NPS and the Forest Service, should be transferred out of the NPS, either into an existing agency with a multiple-use mandate or into a new agency.

—Expand the systems in each country while retaining an emphasis on preservation of diverse lands. All sources, including the renowned President's Commission on Americans Outdoors, indicate that outdoor recreation and wilderness will continue to become "increasingly important" to North Americans. Expansion of available opportunities may be the only realistic way to ease the congestion and overcrowding described in previous chapters. The CPS plan for expansion is well designed and conceived. Expanding it to include maritime areas is a worthy objective. The Canadian government must continue to emphasize representation of diversity and to stipulate natural conditions, despite pressures to allow logging and mining in new parks. The NPS could gain numerous potential areas through acceleration of the review system for the nearly 200 million federal roadless acres not yet designated for wilderness status. Many of these may provide useful buffer zones to help insulate park areas from external threats.

—Remove or reduce the inconsistencies evident in parks that make preservation seem random or sporadic. As Clarke and McCool noted in their comparison of public land agencies, "Agencies that have a strong sense of their own identities and of their mission . . . have an edge over those that do not."[26] Inconsistencies in agency actions give an impression of indecision and confusion.

Numerous actions reflecting inconsistencies by both services were noted in earlier chapters and are mentioned only briefly here. These include such activities as resource extraction, sport hunting, and grazing that one would not expect to see in areas dedicated to preservation. The most prominent inconsistencies, however, are in commercial development. Preservation advocates have encouraged the NPS for years to move more concessions and overnight accommodations outside the parks. Recent efforts such as the plan to remove some concessions from the Yosemite Valley have been overruled. The realization of such efforts will benefit the NPS in the long run. The CPS will probably never remove the town sites from their western parks. As desirable as such a goal may seem to some, minimization of growth and maximization of ecological lifestyles within the towns are more attainable targets. These places will seem less anomalous with strict limits on growth and restrictions on building and zoning.

—Build public consensus behind the preservation goal by involving the public and relevant interest groups in planning efforts. Other studies have shown that public agencies can institute significant changes by involving the public in their planning processes. To some extent, both agencies currently do this, particularly in their management plan reviews. The account of Prince Albert's efforts in chapter 3 illustrates specifically how this can be accomplished. Public awareness of the preservation goal becomes particularly essential as the parks face more external threats over which they have little control.

Both agencies could be more systematic in involving citizens in all phases of planning and encouraging suggestions as to how parks can be enjoyed while still preserved. Some field managers have recognized the potential to build alliances with local groups that support specific goals such as preservation. That support, in a manner counter to Selznick's local-capture argument, can help prevent the subversion of agency goals. In a 1987 address to other park superintendents, Big Bend's Rob Arnberger encouraged his colleagues to build ties in the community that can be a source of later strength.[27]

—Build consensus behind the preservation goal through increased public education efforts. Both agencies currently have some programs intended to inform the public about the importance of and the challenges to parks and historic sites, but neither has been very systematic or sophisticated in its efforts. For example, in 1988 the NPS employed only fifty-six historians. The NPCA Advisory Commission on Resource Management concluded that with "a bold new educational mission . . . park preservation could gain many more advocates."[28]

INCREASING POLITICAL SUPPORT.   Both agencies depend on political support from other relevant actors both inside and outside government. Several procedures could be undertaken to increase levels of support.

—Communicate goals. Simply stating an explicit goal of an agency is not enough; it must be communicated to other relevant participants. Analyses of other agencies, particularly Sonnichsen's study of the Federal Bureau of Investigation, illustrate how communication of an agency's sense of purpose is fundamental to an agency's success.[29] Both the CPS and the NPS could do a better job of communicating the value of their preservation efforts to legislators and potentially supportive interest groups.

In general, neither agency has put much effort into legislative liaisons. Congressional staffers told me the NPS lags badly behind other public land agencies in this regard. Some efficiency experts suggest the fastest way to improve these efforts is through hiring professionals or groups outside of government. Providing timely and accurate information is even more fundamental to an improvement in legislative relations. In this regard, the activities of the Socio-Economic Branch of the CPS and the NPS Statistical Office in Denver are crucial. Both divisions need greater budgetary support and a higher profile in their respective agencies.

—Create more effective presentations for funding. The effort to generate external support could foster immediate gains if used effectively on those who control funding. When revitalizing their agency, FBI employees, for example, successfully revamped their style of presenting budgets to Congress. One emphasis of those efforts should be to eliminate the line-item funding for individual park units that invites the pork-barrel expenditures described earlier and instead seek allocations for the agency to dispense. Short of that perhaps unrealistic goal, pending legislation in the United States would close the loophole in the 1935 Historic Sites Act that allows appropriations for projects not authorized through the normal committee channels. Passage of that legislation would be a good start.

—Strive to broaden political support. The need for greater financial resources in general is a key element in increasing political support. Such a recommendation is admittedly problematic in a time of serious financial deficits, but as one study on government improvement concludes, virtually all "improvements require some new expenditures, an investment."[30] Reduction of budgets for both the NPS and the CPS, when the agencies face increasing challenges and responsibilities, is an invitation to disaster in the long run.

Given the realities of the current deficit situation in both countries, one reasonable step would funnel entrance and user fees directly to the agencies. A system that avoided the general treasury fund would reduce the temptation to make the park services deficit-reducing cash cows. A system that channeled revenue to the agencies rather than to each individual park would negate the potential tendency for any individual park to emphasize raising visitor totals to maximize revenue.

Entrance and user fees in both services should be raised. One simple means for doing this would assess a fee for each person entering a park (perhaps with reduced fares for children) rather than a standard rate for each car. Raising fees for users of the parks is consistent with the preservation focus since "the user's internalization of the full cost of the resource discourages overuse and encourages conservation."[31] The Clinton administration advocates raising entrance fees by roughly 15 percent. Such a dramatic increase will be politically acceptable to park users as long as the greater revenue is used to benefit the parks, not to provide for other federal programs.

Another means of increasing revenue for the NPS is to continue to move in the direction of more realistic concessions contracts, with higher percentages of profits being returned to the agency. Senator Dale Bumpers, Democrat of Arkansas, and the NPCA are currently pushing legislation (S. 1755) to raise fees, funnel revenue back to parks, and make concession contracts competitive.

The CPS could raise leasing fees on residents in the parks. As chapter 3 described, these fees have been historically low and are even more so now as visitation to the parks (and thus potential business for lessees) increases. Again, these revenues should accrue directly to the national parks agency as supplemental and should not replace appropriations.

—Improve prospects for long-term goals. Public agencies, especially when they are pursuing goals like preservation, depend on long-term support. This is often difficult to obtain from politicians operating on a two-year time horizon. Organization specialists recommend building constraints that favor long-term consequences into political relations with agencies. One means for doing so with financial arrangements for the park services is to require permanent allocations. For example, politicians like to allocate initial sums for start-up costs on new projects in the parks and then provide few follow-up resources. Thus the parks' infrastructure, as described earlier, is deteriorating. Long-term funding would be enhanced by explicit requirements in initial appropriations of perennial allocations for maintenance, repair, and replacement.

—Increase political support from officials below the national level. National parks are certainly affected by regional and local behavior, particularly through economic activities and external threats such as air and water pollution. While some efforts are under way, the park services could be much more aggressive in pursuing state and provincial support. Decentralization will help as park officials will then have the authority and prestige with which to deal with state and provincial policymakers. Other agencies have used forums, such as the regional coal teams for BLM, that foster cooperative relations. Individual park managers could involve regional authorities in planning and evaluation processes. Finally, the park services need to do a better job of convincing state and provincial leaders of the benefits that parks provide and explaining how those benefits are threatened by park deterioration. Data such as the material produced by the Socio-Economic Branch of the CPS, described in chapter 4, facilitate such educational outreach efforts.

—Enhance external support for the NPS by removing the agency from the Department of the Interior and providing it with a new institutional home. Others have suggested creating an independent agency. I would encourage instead following the Canadian example and creating a department of the environment in which the agency could be housed. Cabinet status provides needed prestige and access. The placement of several related agencies, such as the Fish and Wildlife Service, in the same department would facilitate the coordination of efforts that is essential for an ecosystem management approach. Attempts to establish an environment department have precedents from both the Nixon and Carter administrations.

The NPS would still be subject to politically appointed leadership and thus susceptible to political direction. During the 1980s, for example, the cabinet leader would have been Anne Gorsuch rather than James Watt, which would have meant perhaps little difference in policy directives. However, the contentious nature of interagency relations in the Department of the Interior would surely be reduced and perhaps replaced by more cooperative efforts.

INSTITUTIONALIZING EMPLOYEE COMMITMENT.   The park services are blessed with a dedicated work force that values the mission of preserving natural areas. Both agencies could institutionalize that commitment by fostering and nurturing it in productive ways.

—Hire individuals with a commitment to agency goals. Kaufman's study of the Forest Service showed how an agency can build commitment by selecting individuals who have the "will and capacity" to pursue agency goals.[32] The hiring process for both park service agencies is somewhat random. Both the NPS and the CPS need to develop more systematic means of hiring people who are committed to and capable of pursuing the preservation goal. Recent changes in the educational requirements for the CPS described in chapter 3 are a step in the right direction. The NPS still awaits political endorsement of similar changes proposed by agency employees as described in chapter 2.

Simply changing educational requirements will not be enough, however. To hire good people, the agencies need an appealing image, perhaps similar to that described by Kaufman for the Forest Service, and enough material rewards to attract qualified individuals. These factors would be particularly beneficial in attempts to hire more employees with backgrounds in science who will be able to perform and use the research that will be necessary to pursue preservation in the future. Both image and compensation are obviously dependent on the consistent, clear mission and on high levels of political support.

—Learn from Experience. In the area of education, the park services could learn from the overall experience of the Forest Service. For years, the FS benefited from the formal schooling and systematic training inherent in the professional forester's degree. In recent years, however, some have argued that such training has become too insulated from the broad background necessary to keep foresters in touch with the demands and goals of the rest of society. John Gordon, dean of the Yale University School of Forestry, has argued that narrow educational training has rendered the FS less effective in incorporating societal needs with scientific management. He urges foresters to pursue a broad liberal arts degree before then going on to two years of professional forestry school.

Ideally, the park services of both countries would benefit from an employee pool with similar training. Future park employees and managers would receive both four years of broad liberal arts education and two years of professional park manager training. This kind of educational training would develop employees with societal awareness and specific skills. Obviously, such efforts require long-term planning and the cooperation of institutions of higher education. The park services could move in this direction fairly quickly, however. A first step would involve the creation of an internship program for undergraduates who are looking for meaningful summertime work with a promising future.

—Enhance employee commitment by appropriate training. Again, Kaufman's analysis is instructive.[33] The orientation sessions, conferences, and training camps of the Forest Service provide a sharp contrast to the occasional, randomly selected courses of the NPS and CPS described in chapters 2 and 3. Employees in both services readily admitted that they would benefit from refresher courses covering subjects ranging from personnel management to fire management. The agencies could require such training for future advancement. So far, such proposals have been inconsistent in the CPS and have been rejected by higher-level authorities in the United States.

Training need not occur only in the classroom. Since so many decisions are specific to each field situation, much training will occur on the job. Field training in critical issues such as scientific resource management is currently quite difficult for a simple reason: the agencies lack the personnel and resources to conduct it. For example, of the 14,000 NPS employees in 1988, only 75 were field scientists. Both agencies need to increase their commitment to providing such training and the resources to make it effective.

—Institutionalize commitment by involving employees in agency goals. Recently many agencies have been using techniques described by Deming and others as Total Quality Management.[34] These techniques prescribe involving employees in every stage of the decisionmaking process of the organization through continuous employee appraisal and communication between managers and employees. Front-line employees offer innovative ideas, and their involvement motivates their later commitment to the organization's efforts.

Such techniques are not necessarily new; they have been used effectively by agencies such as the Forest Service and the Federal Bureau of Investigation. Furthermore, these ideas have been around in both park services for decades. Only recently, however, has the CPS moved seriously in this direction. The regional conferences and task forces described in chapter 3 have already produced innovative ideas and high morale among field employees. Both agencies could foster employee commitment through regular strategy sessions, continuous feedback, and participatory decision-making that cuts across all levels of the organization. These agencies may be more appropriate for techniques involving trail level bureaucrats in decisionmaking than any other agency in either government, since so many actions are unique to each field situation.

—Enhance employee commitment to the preservation goal by emphasizing preservation in advancement decisions. Employees in both the NPS and the CPS currently consider promotion as somewhat random, if not political, with greater appreciation for visitor service and connections than for preservation efforts. Again, other studies have shown that authorities can enhance commitment to specific agency goals by emphasizing performance in relevant efforts in their evaluations. Measuring preservation successes will be a difficult concept, but perhaps the park services could use audits or even outside evaluators to assess progress in mitigating threats to parks or in reversing dangerous trends (for example, revival of endangered species).

—Let those who understand the parks best make the operating decisions in specific locations. Central offices should facilitate system-wide efforts and act as liaisons to other governmental actors. Regional offices can be useful tools, but they should be used primarily for coordination and for dissemination of information, not for control. As other public sector analysts have noted, few reforms could have greater effect on employee morale and commitment than the delegation of real responsibility to enable employees to make decisions in situations they understand best.[35] Once those decisions are made and approved, they warrant support. They could start with the plans already agreed on for bears in Yellowstone, motors in the Grand Canyon, and traffic in Yosemite.

Final Comments

I stayed in the Hoh Forest a long time. After I had hiked out, I realized that I had been so absorbed that I did not get a photograph of the spot where I had seen the green snow. I hiked all the way back in, but of course the moment was not the same. I was reminded of just how elusive even a single event can be. I walked out and drove away, partially satisfied that the moment was burned indelibly into my own memory, even if I could not share it with anyone else. Then, as I crossed over the park's western border, I stopped, again without having planned to. There, on the very boundary of the park, with Mt. Olympus in the distance and the Hoh in the background, was a large, devastated area of clear-cut forest. The ground was covered with charred burn patches and the scattered debris of the recent cut. The jagged, stumpy remnants of trees stood like rough-cut tombstones marking the forest that had existed there before. I recalled a line from Shakespeare that I had seen as the caption on a poster of a ravaged, clear-cut forest area: "O, pardon me, thou bleeding piece of earth, That I am meek and gentle with these butchers!"[36] If we let our national parks suffer a similar fate—cut, paved, dammed or developed for short-term political and economic purposes—we will have lost our own capacity for wonder. This time, I did take a photograph. This was an image that I would share.

Bibliographical Essay

This essay provides a brief summary of sources. I list the interviews separately since they are used freely throughout the work. I then mention sources for each chapter that may not be specified in individual endnotes.

Interviews

Many insights and much information were obtained through interviews with people working for, associated with, or interested in the parks of Canada and the United States. Quite a few of these people preferred to remain anonymous; I have listed the others by country below, with the date of the interview in parentheses.

Interviewees in the United States

Denis P. Galvin, associate director, Planning and Development (March 12, 1991)

Jack Morehead, associate director, Operations (March 28, 1991)

Gene Hester, associate director, Natural Resources (March 13, 1991)

Bill Gregg, coordinator, Man and the Biosphere Program (March 12, 1991)

Barry Mackintosh, National Park Service historian (March 12, 1991)

Anne Frondorf, chief, Planning and Information for Wildlife (March 12, 1991)

Bruce Scheaffer, director of Budget (March 12, 1991, and September 16, 1992)

Rob Arnberger, superintendent, Big Bend National Park (May 31, 1991)

Larry Henderson, superintendent, Guadalupe Mountain National Park (June 4, 1991)

Jan Wobbenhorst, chief ranger, Guadalupe Mountain National Park (June 4, 1991)

Bob Valen, chief naturalist, Guadalupe Mountain National Park (June 4, 1991)

Bill Paleck, superintendent, Saguaro National Monument (June 10, 1991)

John Reed, assistant director, Grand Canyon National Park (June 13, 1991)

Bruce Swadlington, chief, concessions management, Grand Canyon National Park (June 13, 1991)

Ken Miller, chief ranger, Grand Canyon National Park (June 13, 1991)

Dan Cockrum, chief of maintenance, Grand Canyon National Park (June 13, 1991)

Dale Thompson, instructor, Albright (June 13, 1991)

Larry Wiese, superintendent, Zion National Park (June 18, 1991)

George Davidson, chief interpreter, Capitol Reef National Park (June 20, 1991)

Noel Poe, superintendent, Arches National Park (June 21, 1991)

Harvey Wickware, superintendent, Canyonlands National Park (June 21, 1991)

Sheridan Steele, assistant superintendent, Rocky Mountain National Park (June 26, 1991)

Jack Neckels, deputy regional director, Rocky Mountain Region (June 28, 1991)

Paul Pritchard, president, National Parks and Conservation Association (January 2, 1992)

Jerry Schober, former superintendent, Jefferson Memorial (April 3, 1992)

Michael Frome, author (July 16, 1992)

Al Runte, author (July 28, 1992)

Mike Finley, superintendent, Yosemite National Park (August 10, 1992)

B.J. Griffin, director of operations, Western Region (August 14, 1992)

John Sacklin, chief of planning, Yellowstone National Park (August 28, 1992)

Sandy Rives, management assistant, Shenandoah National Park (October 23, 1992)

Interviewees in Canada

Donna Petrachenko, director, Strategic Policy (June 3, 1992)

John Carruthers, policy advisor (June 3, 1992)

Jean-Yves Cayen, policy analyst (June 3, 1992)

Jay Beaman, director, Socio-Economic Branch (June 3, 1992)

Bob Speller, member of Parliament (June 3, 1992)

Luc Perron, analyst, Socio-Economic Branch (June 4, 1992)

Ken James, member of Parliament (June 4, 1992)

Mike Murphy, superintendent, Pukaskwa National Park (June 9, 1992)
Mike Fay, Plains Region office (June 12, 1992)
Richard Leonard, Plains Region office (June 12, 1992)
Alex Zimmerman, Plains Region office (June 12, 1992)
Mac Estabrooks, superintendent, Riding Mountain National Park (June 15, 1992)
Bill Waiser, author (June 16, 1992)
Dave Dalman, superintendent, Prince Albert National Park (June 19, 1992)
Paul Galbraith, chief warden, Prince Albert National Park (June 19, 1992)
Jack Willman, chief warden, Elk Island National Park (June 22, 1992)
Gaby Fortin, superintendent, Jasper National Park (June 25, 1992)
George Andrew, president, Jasper Chamber of Commerce (June 26, 1992)
Perry Jacobson, chief warden, Kootenay National Park (June 30, 1992)
Charley Zinkan, superintendent, Banff National Park (July 2, 1992)
Ian Church, superintendent, Yoho National Park (July 3, 1992)
David Bowes, Western Region office (July 6, 1992)
Jillian Roulet, Western Region office (July 7, 1992)

Note to Chapter 1

The theoretical framework developed in this chapter is based on a variety of works from political science, economics, public administration, and organization theory. Advantages enjoyed by supporters of private over public goods were discussed by Lowi (1979); McConnell (1966); and Schattschneider (1960). Employee motivation is discussed by Clarke and McCool (1985, p. 8); Downs (1967, p. 223); and Heclo (1977, p. 220). The importance of political support is pointed out by Clarke and McCool (1985, p. 5); Rourke (1984, pp. 81–106); and Selznick (1957, p. 21), among others. Much of the discussion of the effect of structure on agency behavior is derived from Kelman (1987); Moe (1989); Selznick (1957); and Warwick (1975). Appointees and careerists are discussed by Aberbach and Rockman (1988); Heclo (1975, 1977); Moe (1989); Pfiffner (1987); and Wilson (1980, 1989).

Literature on the goal of preservation is abundant. For discussion of the ideal of a timeless preserve, see Allin (1987); Frome (1971, p. 141); Hays (1987, pp. 128–29); Junkin (1986, p. 55); and Nelson (1982, p. 32). Belief in this goal as a worthy endeavor is evident in Abbey (1968, pp. 55–63); Herman (1992, pp. 18–36); Lemons and Stout (1984, pp. 49–60); Sax (1980); Shanks (1984, pp. 227–28); and Winks (1993).

The early history of the National Park Service is covered in Everhart (1983); Foresta (1984); and Wirth (1980). Recent problems are discussed in Coates (1991a and b); Freemuth (1991); Frome (1992); Frome, Waver, and

Pritchard (1990); Poindexter (1991); Satchell (1992); and Seligsohn-Bennett (1990). The early history of the Canadian Parks Service is addressed in Brooks (1970); Bryan (1973); Fuller (1970); Lucas (1970); Marsh (1970); Nelson (1970b); Pimlott (1970); and Theberge (1976). More recent changes are covered by Eidsvik and Henwood (1990); Fraser (1979); Nelson (1979); and Stephenson (1983). Data on visitation are available in Canadian Parks Service (1992g); Davidson (1979); U.S. National Park Service, (1992b); and U.S. National Park Service (various years), *Statistical Abstract*.

For more on institutional dissimilarities between Canada and the United States, see Aberbach, Putnam, and Rockman (1981); Cook (1982); Hoberg (forthcoming); Kilgour and Kirsner (1989); Lever (1988); Lipset (1990); Lubin (1986); Nemetz, Stanbury, and Thompson (1986); Russell (1969); Schwartz (1986); Vogel (1993); and Weaver and Rockman (1993). Useful work on Canadian institutions specifically includes Campbell (1983); Franks (1987); Gibbins (1982); Sutherland and Doern (1985); Weaver (1992); and Wilson and Dwivedi (1982). Sources on American institutions include Aberbach and Rockman (1988); Fisher (1991); Savage (1991); Skowronek (1982); and Wood and Waterman (1991). Differences between U.S. and Canadian interest groups are discussed in works by Bella (1987); Doern and Phidd (1983); Hoberg (forthcoming); and Pross (1985). For comparisons of federal structures, see Bakvis and Chandler (1987); Cairns (1992); Feldman and Goldberg (1987); Leman (1987a); and Watts (1991).

The section on recent changes in institutions is based partly on the following. For Canada, see Adie and Thomas (1982); Aucoin (1988); Aucoin and Bakvis (1988); Bakvis (1989); Bercuson, Granatstein, and Young (1986); Campbell (1988); Clarkson and McCall (1990); Harrison and Hoberg (1991); Hoberg (forthcoming); Hoy (1987); Simard (1988); Vogel (1993); and Weaver (1992). For the United States, see Durant (1992); Gottlieb (1989); Kenski and Kenski (1984); Lash (1984); Melnick (1983); Moe (1985); Nathan (1983); Pfiffner (1987); Rosenbaum (1991); Rourke (1991); Vig (1990); and Wilson (1989). For recent changes affecting environmental support, see Clarkson (1982); Hummel (1985); Hunt (1990); Leiss (1979); Magnuson (1990); Mitchell (1990); and Smith (1992).

Note to Chapter 2

Numerous National Park Service (NPS) documents were essential to this chapter. The agency mandate is made explicit in the *National Park Service Organic Act, U.S. Code Annotated*, title 16, section 1. Certain budget figures from the 1980s were obtained and derived from an unpublished document by Denis Galvin, director of planning and development for the

NPS. Threats are described fully in *State of the Parks 1980*. The data were gleaned in large part from U.S. National Park Service *Statistical Abstracts* and U.S. Department of the Interior *Budget Justifications*. One of the most informative and candid documents is U.S. National Park Service *National Parks for the 21st Century* (1992b), an appraisal of the agency on its seventy-fifth anniversary. The *Twelve-Point Plan* (1986) was an internal attempt to lay out agency goals. *Management Policies* (1988c) was used extensively for explicit agency guidelines. The *National Parks Index* (1989b) has useful information on each park.

The background section relied on the following histories: Albright and Cahn (1985); Caulfield (1989); Frome (1992); Hays (1959); Runte (1979, 1987); Shankland (1970); and Sontag (1990). For early agency structure, see Abbey (1968, foreword); Chase (1987b, p. 386); Nienaber and Wildavsky (1973, p. 43); Olsen (1985); and Pyne (1989, p. 114). Work on the tensions surrounding the preservation mandate include Clarke and McCool (1985); Culhane (1978, p. 228); Everhart (1983); Foresta (1984); Hartzog (1988); Nash (1982, p. 253); Nelson (1982); Shanks (1984); and Wirth (1980).

The effects of the Reagan administration on the National Park Service are considered by Cahn and Cahn (1987); Durant (1992, pp. 51–77); Freemuth (1989); Frome (1992); Hartzog (1988); Kraft and Vig (1990); Lash (1984); Lester (1990); and Stanfield (1987). For the Bush administration, see Craig (1991a, 1991b); McManus (1992); Rosenbaum (1991); Turque (1991); Vig (1990); and Worsnop (1993a). The agenda of the wise-use movement is presented in Gottlieb (1989). For the reaction of environmental groups to these effects, see Craig (1991a and b); Foresta (1984, pp. 68–70); Nash (1982); and Pritchard (1992).

In addition to National Park Service documents, a variety of sources discussed individual threats to the parks in recent years. For internal threats, see Adler (1986a and b); Bartimus (1991); Chase (1987a, 1987b, 1991); Coates (1991a and b); Dearden (1985); Dolan (1991); Frome (1992); Garrett (1978); Herman (1992); Huber (1992b); Kaplan (1991); Mitchell (1991); Poindexter (1991); Sax (1982); Segal (1991); Sellars (1989); Shanks (1984); and Wade (1991). For external threats, see William Allen (1991, 1993); Ayres (1991); Barnett (1988); Bodi (1988); Deford (1988); Ferguson (1993); Freemuth (1991); Kenworthy (1991b); McGinley (1988); Seligsohn-Bennett (1990); Shenan and Shabecoff (1988); and Struck (1991). The lack of clarification of agency goals in the face of these threats is discussed by Clarke and McCool (1985, p. 49); Everhart (1983, p. 46); Lemons and Stout (1984, p. 41); and Stegner (1989, p. 44).

The section on diminishing political support benefited from the following sources. Interactions with other agencies are discussed by Bodi (1988); Culhane (1981); Dodd (1988); Gray (1988); Harvey (1988); Hocker

(1988); Keiter (1988); Lockhart (1988); Rauber (1992); Rosenbaum (1985); Sax and Keiter (1988); Shanks (1984); and Zuckerman (1992). For financial considerations, see Bartlett (1985); Clarke and McCool (1985, pp. 59–64); Frome (1992, chapter 10); Herman (1992); Hummel (1987); Lancaster (1991); Lash (1984, pp. 279–83); Phillips (1992); U.S. Congress hearings on entrance fee increases in 1982, 1986, and 1987, and on Department of the Interior appropriations; U.S. General Accounting Office documents on entrance fees (1982) and concessions (1991a, 1991b, 1992a, 1992b). For an expanded version of the entrance fee controversy, see Lowry (1993). For discussion of fees used by other public lands agencies, see Foss (1960); Libecap (1981a, 1981b); Peffer (1951); Rowley (1985); and Voight (1976).

Most of the information on the operating environment of the National Park Service was obtained in interviews. For political assault, also see Cahn and Cahn (1987); Freemuth (1989); Frome (1992); Frome, Waver, and Pritchard (1990, p. 421); Hartzog (1988); Mintzmyer (1992); Poindexter (1991); Rauber (1992); Soden (1991); Stanfield (1987); and virtually every issue of "NPCA News" in the National Parks and Conservation Association (NPCA) magazine *National Parks*. For more on employee morale, see Bartimus (1991); Satchell (1992).

The section on policy behavior uses National Park Service documents such as *Management Policies* (1988c) and *Criteria for Parkland* (1991a). Expansion is also discussed in Adams (1993); Chen (1992); Craig (1991b); Crosson (1991); Epstein (1993); Gilmour (1990); Hinds (1992); Koenig (1991); Lancaster (1990a); National Parks and Conservation Association (1988); Philip (1991); Uhlenbrock (1991); and U.S. General Accounting Office (1991b). Planning is also addressed in Brandborg (1970); Clarke and McCool (1985, p. 57); Fabry (1988); and Heacox (1991). Recreational activities are considered by Edelson (1993); Frome (1992, pp. 197–202); Segal (1991); Shanks (1984, pp. 217–18). Commercial uses of parks are described by Kenworthy (1991b); Masters (1991); Sax and Keiter (1988); and Shanks (1984). For more on the Yellowstone fires and their aftermath, see Freemuth (1989); Frome (1992, p. 170); Malcolm (1989); Monastersky (1988); O'Gara (1989); Pyne (1989); Reid (1989); Schmidt (1989); Schwartz (1992); Shabecoff (1988); Sholly and Newman (1991); Suro (1990); U.S. Congress, Senate Subcommittee on Public Lands, National Parks, and Forests Hearings (1988); U.S. National Park Service (1991d and 1992d); and Wuerthner (1989). Science and education are discussed in Ayres (1988); Chase (1987); Deford (1988); Kenworthy (1992c); Shenan and Shabecoff (1988); and Twomey (1992). On employee dissatisfaction, see Cahn and Cahn (1987); Frome (1992, p. 141); Hartzog (1988: 271–72); Rauber (1992); Shanks (1984); and Stanfield (1987). For budgets, see Allen (1993); Hinds (1992); Kanamine (1993); Lancaster (1990a); Meier (1993); Newman

(1993); National Parks and Conservation Association (1992c and 1993b); U.S. General Accounting Office (1991b); and Wade (1993).

Note to Chapter 3

I used a variety of government documents for this chapter. The Canadian Parks Service (CPS) document *State of the Parks* (1990e) is full of information. The CPS document *Proposed Policy* (1991b) is used extensively in the latter part of the chapter. I used Canadian Parks Service documents such as *A Bridge to the Future* (1992a); *Revenue Profile* (1991c) and *Visitor Participation Statistics* (1992g); Beaman and Stanley (1991); and Environment Canada, *Report on Transition* (1991). Environment Canada documents such as *Legislation Governing the National Parks of Canada* (1988) and *Canada's Green Plan* (1990) were helpful.

For more on the early history of the agency, see Brown (1970); Bryan (1973); Cahn and Cahn (1992); Canadian Nature Federation (1981); Eidsvik and Henwood (1990); Geist (1979); Lothian (1987); Lucas (1970); Nelson (1970a); Stewart and Finkelstein (1985). Descriptions of early commercial activity include Davis (1989); Begin (1989); Bella (1987); Brown (1970); Bryan (1973); Clarke (1979); Fuller (1970); Nelson (1970b); Nicol (1970); Nikiforuk (1990); and Theberge (1976). For recent emphases on forestry and energy self-sufficiency, see Barney, Freeman, and Ulinski (1981); Clarkson (1982); and Postel and Ryan (1991). Discussions of early town sites, particularly Banff, are available in Bella (1987, pp. 113–15); Leighton (1985); and Scace (1968). For changes during the 1970s, see Brooks (1970); Burton (1979); Carruthers (1979); Davidson (1979); Hummel (1985); Leiss (1979); and Marsh (1979).

Topics less specific than the parks were covered in the following works. Canadian traditions of deference to political authorities are discussed in Aberbach, Putnam, and Rockman (1981); Campbell (1983, p. 296); Cook (1982, p. 12); and Lipset (1990, p. 212). For more on the importance of provincial governments, see Bryan (1973, pp. 248–50); Cairns (1992, p. 65); Eidsvik and Henwood (1990, p. 65); Hoberg (forthcoming, p. 20); Leman (1987a, pp. 29–30); Lewis (1979, p. 69); Nicol (1970, pp. 19–25); Taylor and Nixon (1990, p. 25); Watts (1991); and Weaver (1992, pp. 43–44).

My account of changes in the political environment is based largely on news reports and the following works: Begin (1989); Bella (1987); Bercuson, Granatstein, and Young (1986); Bergman (1986); Borins (1988); Campbell (1988); Clarkson and McCall (1990); Gwyn (1980); Harrison and Hoberg (1991); Hoberg (forthcoming); Hoy (1987); Howse (1989);

Nikiforuk (1990); and Weaver (1992). For structural changes, see Adie and Thomas (1982); Aucoin and Bakvis (1988); Sutherland and Doern (1985); and Wilson and Dwivedi (1982). Journalistic accounts of threats include those by Glen Allen (1991); Aniskowicz (1990); Bohn (1985); Bohn and Barrett (1987); Carp (1989); Dawson (1990); Dearden (1985); Hill (1985); Hume (1990); Leighton (1985); Masterman (1989); McNamee (1992); Nudds and Thomas (1990); Ovenden (1990b); Patterson (1983); Platiel (1983); Struzik (1990); Werier (1991); and Winning (1982). Increased environmental awareness is discussed by Barrett and Huff (1986); Barnett (1992); Claiborne (1990); Canadian Nature Federation (1991); Huff (1985); Hume (1988); Hummel (1985); Lowey (1985); Walker (1992); and Wren (1985).

As with the NPS, much of my analysis of changes in the operating environment of the CPS derives from interviews with agency personnel. Also see Canadian Parks Service regional and task force documents such as *Choosing Our Destiny* (1990a); *Environment Canada Integration Task Force* (1991a); and *Report of the Science and Protection Task Force* (1990d). For reference to structural changes in a broader context, see Aucoin and Bakvis (1988, p. 42); Chodos, Murphy, and Hamovitch (1988, pp. 27–30); Crozier (1964, p. 189); Hoy (1987, p. 124); and Sawatsky (1991, pp. 513–14).

Much of the section on policy change is based on the CPS documents *National Parks System Plan* (1990c), *Proposed Policy* (1991b), and *State of the Parks* (1990e). For reference to system expansion, see Brue (1989); Bryden (1987); Cahn and Cahn (1992); Carruthers (1979); Canadian Environmental Advisory Council (1991); Davidson (1979); Eidsvik and Henwood (1990); Huff (1985); Israelson (1989); Lowey (1987); McLaren (1986); Munro (1987); Nikiforuk (1990); and Walker (1991 and 1992). For more on management policies, see Bella (1987); Church (1988b); Canadian Nature Federation (1979); Cox (1987); Eidsvik and Henwood (1990); Galbraith (1989); Graham, Nilsen, and Payne (1987); Leighton (1985); Nelson (1984); Ovenden (1992); Purser (1982); Stephenson (1983); Taylor (1990); Theberge (1981); Val (1987); and Van Wagner and Methuen (1980). Data on the use of resources can be found in Environment Canada (1992); Canadian Parks Service, *Final Report of the Parks and People Task Force* (1990); and other CPS documents.

Note to Chapter 4

Most of the information for this chapter about field programs came from going to the parks in question (a tough job, but somebody had to do it), seeing the problems, and talking to the people who work there. In addition, I used the following sources.

Recent books on Yellowstone include Bartlett (1985); Chase (1987b); Ferguson (1993); Schullery (1986); and Sholly and Newman (1991). National Park Service documents include park brochures, the *Final Environmental Impact Statement* for Yellowstone National Park (1988a); the *Yellowstone Master Plan* (1974); the *Yellowstone Statement for Management* (1991); and *State of the Rocky Mountain Region* (1990b). See also Craighead and Craighead (1971); Freemuth (1989); Frome (1992, p. 187); Glick (1992); Kenworthy (1992a); Linden (1992); McNamee (1984); Rauber (1992); Schnoes and Starkey (1978); Seligsohn-Bennett (1990); and Struck (1991).

Documents with information on Elk Island include Canadian Parks Service *State of the Parks* (1990e); *Elk Island National Park Management Plan Review* (1992d); and *Application Design Report Ecosystem Management Model Project* (1992b).

For more on Grand Canyon controversies, see Abbey (1981); Carothers and Brown (1991); Crumbo (1981); Fradkin (1981); Garrett (1978); Ghiglieri (1992); Nash (1982, pp. 210–17); and Sax (1980, pp. 91–93). Specifically related to boat and air use of the canyon are Borden (1976); Cahn and Cahn (1987); Coates (1991a); Everhart (1983, pp. 102–05); Frome (1992, p. 44); Hartzog (1988, p. 9); Parent and Robeson (1976); Shanks (1984, p. 225); Stanfield (1987); Shelby and Nielsen (1976); Thompson, Rogers, and Borden (1974); and Van der Wert and Barber (1992). Documents include U.S. National Park Service, *Colorado River Management Plan* (1979a, 1989a); *Final Environmental Impact Statement of Proposed Colorado River Management Plan* (1979b); and *Grand Canyon Natural Resources Management Plan and Environmental Assessment* (1977a).

Jasper controversies are discussed in Editors of the *Calgary Herald* (1984); Harasymuk (1991); Kaegi (1992); Leeson (1979); Leighton (1985); and Masterman (1991). CPS documents include *Banff National Park Management Plan* (1989a); *Jasper National Park Management Plan* (1988b); *Private-Sector Construction* (1989b); *Revenue Profile* (1991c); *State of the Parks* (1990e).

Yosemite issues are discussed in Brower (1990); Diringer (1989 and 1991); Everhart (1983); Frome (1992); Hatfield (1992); Herman (1992); Huber (1992a); Jones (1989); Muir (1912); Nolte (1990a, b, and c); Philip and Huber (1991); Phillips (1992); and, especially, Runte (1990). NPS documents include "Yosemite Valley/El Portal Comprehensive Design" (1987b) and *Yosemite General Management Plan* (1980d). See also U.S. Congress, Senate Subcommittee on Public Lands, National Parks, and Forests Hearings (1990); U.S. Department of the Interior (1990a); and U.S. General Accounting Office (1992a).

For the early history of Banff, see Baird (1977); Lothian (1987); Luxton (1975); Nelson (1970); Parks Canada (1979, p. 45); and Scace (1968). For more recent controversies, see Bella (1986); Canadian Nature Federation (1980); Geddes (1989); Howse (1989); Leighton (1985); Marty (1984,

1985); Masterman (1989); Nelson (1984); Ovenden (1992); and Patterson (1979). See also Canadian Parks Service (1988a and 1991c, p. 70).

Government documents discussing Shenandoah include U.S. National Park Service, "Air Quality Fact Sheet" (1992a), *Shenandoah General Management Plan* (1983), and *Second Annual Shenandoah Research Symposium* (1977b); and U.S. Congress, House Subcommittee on Environment, Energy, and Natural Resources Hearings (1990). See also Ayres (1991); Cahn and Cahn (1987); Cohn (1990a, b); Conners (1988); Diglio (1989); Freemuth (1991); Harris (1991); Masters (1991); and PEI Associates (1989).

CPS documents on Prince Albert include *Prince Albert National Park Management Plan Summary* (1987); *State of the Parks* (1990e); *The Fires of Change* (1992e); *Waskesiu Community Plan* (1988c); and *Wolf Country* (1992h). See also BOREAS (1991); Rowe (1990); and, for a very informative study of the park, Waiser (1989).

Note to Chapter 5

Most of the sources used in the summary section have already been used elsewhere in the work. For summary comments on the state of the NPS, see Clarke and McCool (1985); Conservation Foundation (1985); Frome (1992); Soden (1991); and U.S. National Park Service (1992b). Sources for the section on the Clinton administration include Cockburn (1993); Cohn and Williams (1993); Kamen (1993); National Parks and Conservation Association (1993a); Taylor (1993); Winks (1993); and Worsnop (1993a and b). Summary assessments of the CPS are in Bella (1987); Cahn and Cahn (1992); Canadian Nature Federation (1991); Eidsvik and Henwood (1990); Huff (1985); Nikiforuk (1990); Stephenson (1983); and Minister of the Environment's Task Force on Park Establishment (1987).

Various theoretical contributions were used in the section on parks in context. The importance of the political environment on other agencies in similar contexts is discussed by Halperin (1974, p. 51); Lowi (1979, p. 93); Mazmanian and Nienaber (1979, p. 59); Moe (1989, p. 324); Selznick (1957, p. 21); and Wilson (1989, p. 26). For effects of the operating environment, see Aucoin and Bakvis (1988); Durant (1992); Hayek (1948); Knott and Miller (1987); Landau and Eagle (1981); Moe (1985); Nathan (1983); Reich (1985); Rourke (1991); and Wilson (1989). For more on decentralized agencies, see Aucoin and Bakvis (1988); Kaufman (1960); Reich (1985); and Selznick (1949). The effect of fee increases is discussed by Lowry in article form (1993).

Application of concepts to other agencies necessitated the use of a variety of other sources. For more on the Grazing Service and the Bureau of Land Management, see Clarke and McCool (1985, pp. 110–11); Culhane

(1981, pp. 81–91); Durant (1992, pp. 21–22); Foss (1960); Gates (1968, p. 619); Libecap (1981b); Peffer (1951); and Shanks (1984, pp. 181–82). For analyses of the Corps of Engineers, see Anderson and Binstein (1992); Clarke and McCool (1985, pp. 18–28); Ferejohn (1974); Maass (1951); Mazmanian and Nienaber (1979); Reisner (1986); and Shanks (1984, pp. 131–51). The classic on the Forest Service is, of course, Kaufman (1960). For more, see Clarke and McCool (1985, pp. 37–42); Culhane (1981); Durant (1992, p. 266); Egan (1990); Leman (1987b); Rauber (1991); Riley (1992); Scott (1991); Shanks (1984, pp. 153–71); Tipple and Wellman (1991); and Zuckerman (1992).

The section on recommendations used sources from several different literatures. Those encouraging more reliance on market forces in the parks include Anderson and Hill (1981); Gottlieb (1989); Johnson (1981); and Stroup (1990). The difficulties of doing so are argued by Dorfman (1993, p. 91); Kelman (1981, p. 40); Kremp (1981, p. 147); Krutilla (1967, p. 191); Nelson (1982, p. 66); and Solow (1991, p. 182). Proposals to restructure political institutions governing parks are discussed by Fairfax and Yale (1987); Francis and Ganzel (1984); and Nelson (1982). Those who encourage making the NPS an independent agency include Lockhart (1988); Mintzmyer (1992); and Pritchard (1991a). Others who argue for changing the behavior of the agencies include Frome (1992, pp. 230–31); Leman (1984, pp. 121–24); and Nelson (1982, pp. 53–64). The need for an explicit mission is recognized by Clarke and McCool (1985, p. 7); Downs and Larkey (1986, p. 220); Durant (1992, p. 238); Foresta (1984, p. 104); Lipsky (1980, p. 199); and Nelson (1982, p. 53). The need to reorganize public lands in the United States is mentioned by Cordell (1990, p. 246); Davis and Davis (1987); Junkin (1986, p. 68); and Nelson (1982, p. 59). Public involvement in planning for other agencies is described by Leman (1984, p. 123); and Mazmanian and Nienaber (1979, pp. 27–32). Agency communication skills are discussed in Downs and Larkey (1986); Sonnichsen (1987); and Wholey (1987). For more on efforts to create an environment department in the United States, see Clarke and McCool (1985, p. 25); Gordon (1989, p. 17); Russell (1990, p. 76); and Whitaker (1976). Means to improve worker training are described by Gordon (1984, pp. 62–63); Kaufman (1960, pp. 161–70); and Leman (1984, p. 124). Socialization of employees is analyzed by Cohen and Brand (1993); Durant (1992); Kaufman (1960); and Sonnichsen (1987). Evaluations are described by Durant (1992, p. 272); and Wholey (1987, p. 183) For more on decentralization, see Hayek (1948, p. 79); Kelman (1987, p. 282); and Lipsky (1980, p. 199).

Notes

Chapter 1

1. Quoted in Healey (1992, p. 23).
2. Quoted in Healey (1992, p. 21).
3. Wilson (1989, p. 182).
4. *National Park Service Organic Act. U.S. Code Annotated,* title 16, sec. 1: 66.
5. *National Parks Act of Canada,* 1930.
6. U.S. National Park Service (1988c, p. 1.1).
7. The language of the March 1, 1872, act of Congress is repeated in U.S. National Park Service (1991d, p. 5).
8. Sellars (1992, p. C5).
9. Stroup (1990, p. 169).
10. Survey by Daniel McCool cited in Foresta (1984, p. 104).
11. Lemons and Stout (1984, p. 50); also Herman (1992, pp. 25–36).
12. Sax (1980).
13. Quoted in Coates (1991b, p. 16).
14. Rowntree and Orr (1978, p. 405).
15. U.S. National Park Service (1992b, p. 12).
16. Bryan (1973, p. 293).
17. Eidsvik and Henwood (1990, p. 79).
18. Kaufman (1960, p. 161).
19. Halperin (1974, p. 51).
20. Ripley and Franklin (1976, p. 48).
21. Wilson (1989, p. 181).
22. Rosenbaum (1994, p. 123).
23. Pressman and Wildavsky (1984, pp. 70, 90); Lowi (1979, p. 93).
24. Rosenbaum (1989, p. 225).
25. Clarke and McCool (1985, p. 7).

26. Hayek (1948, p. 79) encouraged the allocation of decisions to levels where "we can expect that fuller use will be made of the existing knowledge."

27. Lipsky (1980, p. 199).

28. Kelman (1987, p. 282).

29. Moe (1989, p. 268). See also Noll (1985, p. 12); Wilson (1989, p. 23).

30. Katzmann (1981, p. 7).

31. Campbell (1983, p. 295); Franks (1987, p. 98).

32. Kelman (1987, p. 91).

33. Heclo (1977, pp. 191–234); Skowronek (1982, p. 287); Weaver and Rockman (1993, p. 32); Wood and Waterman (1991).

34. Campbell (1988, p. 315); Cook (1982, p. 12); Harrison and Hoberg (1991, p. 22); Kelman (1987, p. 91); Lipset (1990, p. 49).

35. Doern and Phidd (1983, p. 31).

36. Hoberg (forthcoming).

37. Cairns (1992, p. 55).

38. Tellier (1990, p. 124).

39. Moe (1985, p. 268); Nathan (1983, p. 7); Rourke (1991, p. 133).

40. Cook (1982, p. 12).

41. Lipset (1990, p. 3).

42. Melnick (1983, pp. 3, 10–12); Wilson (1989, p. 279).

43. Lowi (1979, p. 107). See also McConnell (1966).

44. Sutherland and Doern (1985, p. 174). See also Campbell (1988, p. 309); Wilson and Dwivedi (1982, p. 11).

45. Canadian Government (1990).

46. Robert Cameron Mitchell (1990, p. 81).

47. Dunlap, Gallup, and Gallup (1992, p. 43).

Chapter 2

1. Runte (1979, p. 97).

2. Frome (1992, p. 8).

3. Shankland (1970, p. 4).

4. Albright and Cahn (1985, p. 16).

5. U.S. Congress, *Congressional Record* (1916, p. 10364).

6. Albright and Cahn (1985, p. 36).

7. U.S. Congress (1916, *Congressional Record*, pp. 13004–05).

8. Abbey (1968, foreword); Pyne (1989, p. 114); Chase (1987b, p. 386); Nienaber and Wildavsky (1973, p. 43).

9. Frome (1971, p. 141); Everhart (1983, p. 180); Culhane (1978, p. 238).

10. Clarke and McCool (1985, pp. 48–64).

11. Clarke and McCool (1985, p. 59).

12. Watt's remark was reported by Omang (1981, p. A9); Mott was quoted by McCombs (1985, pp. D1–D2).

13. Soden (1991, p. 574).

14. Stanfield (1987, p. 318).

15. Gottlieb (1989, p. 10).

16. National Parks and Conservation Association solicitation, Spring, 1992.

17. Shute (1992, p. 27).

18. Mackintosh (1984, p. 63).

19. U.S. General Accounting Office (1987, p. 4). See also Magraw (1988).

20. U.S. Department of the Interior (1926, p. 1).

21. The 1917 figure is from U.S. Department of the Interior (1926, p. 1). The 1960 figure is found in U.S. National Park Service (1980b, p. 34). The 1970 figure is from U.S. National Park Service (1980c, table 2).

22. Editors of *The Economist* (1993, p. 31).

23. Devoto (1953, p. 51).

24. Coates (1991b, p. 1). See also Garrett (1978); Adler (1986b); and Poindexter (1991).

25. Coates (1991b, p. 16).

26. Chase (1987b, p. 233). Nevertheless, when I asked about Chase's book at the Old Faithful bookstore, an elderly saleswoman pointed it out to me with a friendly, "Oh yes, Honey, that's a good one."

27. U.S. National Park Service (1980b, p. viii).

28. U.S. National Park Service (1980b, p. 21).

29. Lemons and Stout (1984, p. 41).

30. U.S. National Park Service (1992b, p. 14).

31. U.S. National Park Service (1992b, p. 12).

32. U.S. Department of the Interior (1992).

33. Shanks (1984, p. 104).

34. For a history of preceding secretaries, see Frome (1992, chapter 3).

35. Lockhart (1988, p. 18).

36. Lash (1984, p. 289).

37. Rauber (1992, p. 26).

38. U.S. National Park Service (1986, p. 4).

39. See Watt's testimony in U.S. Congress, Senate Committee on Energy and Natural Resources (1982, p. 221).

40. U.S. General Accounting Office (1982, p. 8); U.S. Congress, Senate Subcommittee on Public Lands, National Parks, and Forests (1987, p. 35).

41. U.S. Congress, Senate Subcommittee on Public Lands, National Parks, and Forests (1986, p. 82) and (1987, pp. 52, 53).

42. See NPS Director Mott's testimony in 1986 in U.S. Congress, Senate Subcommittee on Public Lands, National Parks, and Forests (1986, p. 93; see also pp. 104, 142, 153, 160) and Horn's testimony in 1987 in U.S. Congress, Senate Subcommittee on Public Lands, National Parks, and Forests (1987, p. 50).

43. U.S. Congress, Senate Subcommittee on Public Lands, National Parks, and Forests (1986, p. 82). See the testimony of Katy Miller Johnson in U.S. Congress, Senate Subcommittee on Public Lands, National Parks, and Forests (1987, p. 81; see also p. 148) and U.S. Congress, *Congressional Record* (1987, April 1, p. H1722).

44. U.S. Congress, Senate Subcommittee on Public Lands, National Parks, and Forests (1987, pp. 83, 148); U.S. Congress, Senate Committee on Energy and Natural Resources (1982, pp. 221–22); Albright and Cahn (1985, p. 37).

45. U.S. Congress, Senate Subcommittee on Public Lands, National Parks, and Forests (1987, p. 85).

46. U.S. Senate, Senate Subcommittee on Public Lands, National Parks, and Forests (1986, p. 92) and (1987, pp. 85, 32, 27, 29, 31).

47. Public Law 100-203 (c) (3).

48. U.S. Department of the Interior, (1990a, p. NPS-12).

49. U.S. Congress, Conference Report on Appropriations (1986, p. 20).

50. *Concessions Policy Act of 1965*, 16 U.S.C. & 20. Also see Herman (1992).

51. U.S. General Accounting Office (1991a, p. 4).

52. U.S. General Accounting Office (1991a, p. 8).

53. Editors of *USA Today* (1992, p. A6).

54. U.S. General Accounting Office (1991a, p. 14); Editors of *USA Today* (1992, p. A6); Associated Press (1992, p. A4).

55. Lancaster (1991, p. A1); Phillips (1992, p. A10); U.S. General Accounting Office (1992b, p. 3); National Parks and Conservation Association (1993c, p. 13).

56. U.S. General Accounting Office (1992a); Fultz (1992).

57. Lancaster (1990b, p. A19).

58. Hartzog (1988, p. xvii).

59. Ripley and Franklin (1976, p. 5) .

60. Hartzog (1988, p. 272). See also Cahn and Cahn (1987, pp. 32–33); Frome (1992, pp. 104–05; "This Week with David Brinkley," July 7, 1991.

61. Frome (1992, p. 12). Also see Cahn and Cahn (1987, pp. 32–33).

62. National Parks and Conservation Association (1991a, p. 11); Rauber (1992); also Special Report by Larry Lamont on CNN (July 28, 1992); National Parks and Conservation Association (1992b, p. 8); Mintzmyer (1992, p. 25).

63. Barr (1992, p. A11).

64. U.S. Congress, House Committee on Appropriations (1983, p. 7); (1984, p. 8); (1985, p. 25); (1986, p. 21); (1987, p. 26); (1988, p. 19).

65. Shanks (1984, p. 220). See also Bartlett (1985); Chase (1987b, chapter 13); Frome (1992, chapter 10). Quoted in Herman (1992, p. 39). U.S. Department of the Interior (1990b, p. 13); Herman (1992, p. 39).

66. National Parks and Conservation Association (1993c, p. 13); Phillips (1992, p. A10); U.S. General Accounting Office (1991a, p. 8).

67. McAllister (1991a, p. A21); Ybarra (1990, p. A6); comments by Senator Wallop in U.S. Congress, Senate Subcommittee on Public Lands, National Parks, and Forests Hearings (1990, p. 115). Quoted in Herman (1992, p. 51).

68. U.S. General Accounting Office (1987, p. 4).

69. U.S. National Park Service (1991b, p. 3).

70. U.S. National Park Service (1992b, p. 29).

71. U.S. National Park Service (1992b, p. 26).

72. Freemuth (1989, pp. 284–85).

73. Soden (1991, p. 571).

74. Aucoin and Bakvis (1988, p. 89).

75. Conservation Foundation (1985, p. 300).

76. Frome, Waver, and Pritchard (1990, p. 421).

77. Chase (1991, p. 40).

78. U.S. National Park Service (1992b, p. 26).

79. Devoto (1953, p. 50); Clarke and McCool (1985, p. 58).

80. Heclo (1977, p. 103); Hartzog (1988, p. 262).

81. Survey conducted by Daniel McCool, cited in Foresta (1984, p. 104); Cahn and Cahn (1987, pp. 32–33); Frome (1992, pp. 100, 104–05).

82. U.S. Department of the Interior (1990a, p. NPS-259); National Parks and Conservation Association (1992e p. 11).

83. U.S. National Park Service (1992b, pp. 9, 47).

84. U.S. National Park Service (1992b, p. 104).

85. U.S. National Park Service (1991c, p. 1).

86. U.S. National Park Service (1988c, p. 2.4).

87. Swem (1970, p. 252); U.S. National Park Service (1972). See also National Parks and Conservation Association (1988, p. 34).

88. CQ Press (1979, pp. 704–05); Gilmour (1990, p. 98).

89. As reported by Omang (1980, p. A5).

90. U.S. Congress, Senate Committee on Energy and Natural Resources (1981, p. 31).

91. Actual percentages are 55 percent federal in Biscayne and 26 percent federal in Channel Islands. Calculated from U.S. National Park Service (1989b, pp. 25, 34).

92. Reid (1987, pp. A1, A11). Editors of the *Washington Post* (1987, p. A22).

93. U.S. National Park Service (1992b, p. 115).

94. Hinds (1992, p. 12); cited in Lancaster (1990a, p. A8).

95. National Parks and Conservation Association (1988, p. 34).

96. U.S. National Park Service (1992b, p. 135).

97. Cohn (1992, pp. B1, B5). Wolf was quoted in Hall (1992, p. D3). See also Simon (1992, p. C8).

98. Fabry (1988, pp. 92–94); U.S. National Park Service (1976b, p. 8); U.S. National Park Service (1980a, p. 18).

99. U.S. National Park Service (1988c, pp. 2.7, 2.8).

100. U.S. National Park Service (1988c, pp. 2.6, 2.7).

101. U.S. Congress, House Committee on Interior and Insular Affairs (1992).

102. U.S. National Park Service (1988c, p. 8.1).

103. Calculated from figures in U.S. Department of the Interior (1990a, p. NPS-12). See also U.S. National Park Service (1992b, pp. 70–71).

104. Ridenour (1991, p. 22).

105. "This Week with David Brinkley," July 7, 1991; Ben Brown (1991, p. 12C).

106. McAllister (1991b, p. A21); Kenworthy (1991b, p. A4).

107. U.S. National Park Service (1988c, p. 8.10); U.S. Congress, *Congressional Record* (1916, p. 13005).

108. Kenworthy (1991c, p. A3; and 1991a, p. A4); National Parks and Conservation Association (1992a, pp. 8–9).

109. U.S. National Park Service (1988c, p. 8.8).

110. Quoted in O'Gara (1989, p. 46).

111. U.S. National Park Service (1988c, p. 4.14).

112. U.S. National Park Service (1991d, p. 33).

113. U.S. National Park Service (1974, p. 24).

114. Chase (1987b, pp. 92–97); Freemuth (1989, p. 283); Pyne (1989, p. 86); Shabecoff (1988, p. A22); Wuerthner (1989, p. 43).

115. Sellars (1989, p. B1); Schmidt (1989, p. 26).

116. U.S. Congress, Senate Subcommittee on Public Lands, National Parks, and Forests (1988, p. 13). Also see O'Gara (1989, p. 44); Shabecoff (1988, p. A22).

117. Quoted in Monastersky (1988, p. 190).

118. U.S. Congress, Senate Subcommittee on Public Lands, National Parks, and Forests (1988, p. 11).

119. Interagency agreement included in U.S. National Park Service (1992d).

120. Reprinted in U.S. National Park Service (1992d).

121. *Washington Post* (1988, p. 12).

122. U.S. National Park Service (1992d, pp. 18–20). U.S. National Park Service (1991d, p. 33).

123. Pyne (1989, pp. 145, 111).

124. Quoted in Reid (1989, p. A3).

125. Chase (1987b, pp. 237–39, 252–53); National Parks and Conservation Association (1988, p. 7); Kenworthy (1991a, p. A4); National Parks and Conservation Association (1992a, pp. 8–9).

126. U.S. National Park Service (1988c, p. 4.2).

127. U.S. National Park Service (1988d, pp. 4–5, 31); (1990a, p. 1); (1988c, p. 4.3).

128. Gordon (1989, p. 16); Russell (1990, p. 76); Cahn (1993, pp. 25–27); National Parks and Conservation Association (1992f, p. 15); U.S. National Park Service (1992b, p. 31).

129. Kenworthy (1992c, p. A25); Twomey (1992, pp. D1, D5).

130. U.S. National Park Service (1988c, p. 7.1).

131. U.S. National Park Service (1992b, pp. 24, 47).

132. Gordon (1989, p. 16).

133. U.S. National Park Service (1992b, pp. 101, 121; and 1980b).

134. U.S. National Park Service (1992b, p. 48).

135. U.S. National Park Service (1992b, p. 54).

136. National Parks and Conservation Association (1988, p. 42).

137. U.S. National Park Service (1992b, p. 58).

138. Chase (1987b, pp. 259–60).

139. U.S. National Park Service (1987a, charts 1 and 2).

140. National Parks and Conservation Association (1991b, p. 18).

141. National Parks and Conservation Association (1988, p. 9). Also see Chase (1991, p. 34).

142. Chase (1987a, p. 43).

143. U.S. General Accounting Office (1991b, p. 1); U.S. Department of the Interior (1990a).

144. U.S. National Park Service (1992c, p. 1).

145. Lancaster (1990a, p. A1).
146. National Parks and Conservation Association (1992d, p. 12).
147. Meier (1993, p. 12); New York Times News Service (1993, p. 9A).

Chapter 3

1. Theberge (1976, pp. 194–95).
2. Eidsvik and Henwood (1990, p. 75).
3. Bella (1987, p. 2).
4. John Carruthers, in an interview on June 3, 1992.
5. Quoted in Eidsvik and Henwood (1990, p. 62).
6. Doerr (1938, p. 5–7).
7. Cragg (1970, p. 113).
8. Theberge (1976, p. 195).
9. Bryan (1973, p. 253).
10. Nicol (1970, p. 21).
11. Davidson (1979, p. 26).
12. Davidson (1979, p. 23).
13. Bryan (1973, p. 263).
14. Reuters News Service (1966, p. 54).
15. Carruthers (1979, p. 656).
16. Monte Hummel (1985, p. 585).
17. Canadian Parks Service (1990e, p. 24).
18. Leeson (1979).
19. Marsh (1982, p. 65).
20. Canadian Assembly on National Parks (1986, p. 124).
21. Canadian Parks Service (1990, p. 191).
22. J. G. Nelson (1979, p. 18).
23. Nikiforuk (1990, p. 37); Ovenden (1990, p. B5).
24. Ovenden (1990, p. B5).
25. Fraser (1985, p. 6).
26. Nikiforuk (1990, p. 31).
27. Werier (1991, p. 6).
28. Dawson (1990, p. B6); Winning (1982, p. C1).
29. Canadian Parks Service (1990e, p. 29).
30. Hummel (1985, p. 586).
31. Gwyn (1980, p. 175).
32. See Trudeau's speech to the Liberal party of Vancouver (May 1, 1971) in Trudeau (1972, p. 134) and Bella (1987, p. 115).
33. From Canadian Parliament, House of Commons (1983, p. 24465).
34. Bercuson, Granatstein, and Young (1986, p. 6).
35. Hoberg (forthcoming).
36. Canadian Parks Service (1991b, p. 14).
37. Bill Waiser, in an interview on June 16, 1992.

38. Lamb (1991, p. B2).
39. Nelson and Patterson (1979, p. A1).
40. Pross (1985, p. 246).
41. Hoberg (forthcoming).
42. Wilson and Dwivedi (1982, p. 4). For the roots of this, see Cole (1966, pp. 23–28).
43. Environment Canada (1988, p. 20).
44. Franks (1987, p. 257); Pross (1985, pp. 235–40).
45. Franks (1987, p. 98).
46. Hoberg (forthcoming).
47. Bergman (1986, p. 115).
48. Aucoin (1988, p. 354).
49. Canadian Parks Service (1990a, p. 4).
50. Environment Canada (1991, pp. 12–13).
51. Crozier (1964, p. 189). See also Aucoin and Bakvis (1988, p. 98).
52. Parks Canada (1986, p. 2).
53. Wilson (1989, p. 153). See also Crozier (1964, p. 304) and Gouldner (1954).
54. Seale (1979, p. A7).
55. Parks Canada (1979, p. 37); Environment Canada (1988, p. 15).
56. Carruthers (1979, p. 648).
57. Canadian Parks Service (1990c, p. 7).
58. Canadian Parks Service (1990e, p. 79); Environment Canada (1990, p. 13).
59. Canadian Environmental Advisory Council (1991, p. 23).
60. Canadian Parks Service (1990c, p. 9).
61. Rowe (1990, pp. 31–33).
62. Canadian Parks Service (1992f, p. 4).
63. Canadian Parks Service (1991b, p. 31).
64. Parks Canada (1979, p. 40).
65. Canadian Parks Service (1991b, p. 32). See also Canadian Nature Federation (1979, p. 21); Theberge (1981, p. 23).
66. Canadian Parks Service (1991b, pp. 31–32).
67. Canadian Parks Service (1992f, p. 4).
68. Canadian Parks Service (1991b, p. 35).
69. Graham, Nilsen, and Payne (1987); Val (1987).
70. Canadian Parks Service (1989a, pp. 1–2, 9–10, 16).
71. Galbraith (1989, p. 51).
72. Canadian Parks Service (1991b, p. 35).
73. Canadian Press (1986, p. 14); Canadian Parks Service (1991b, p. 37).
74. Parks Canada (1979, p. 45). See also Leighton (1985, p. 20).
75. Canadian Parks Service (1991b, p. 9).
76. Canadian Parks Service (1991b, p. 38).
77. Quoted in Canadian Press (1986, p. 14).
78. Canadian Parks Service (1992f, p. 10).
79. Canadian Parks Service (1991b, p. 30).
80. Canadian Parks Service (1991b, p. 37).

81. Canadian Parks Service (1991b, pp. 30–31).
82. Church (1988a, p. A12).
83. Parks Canada (1979, p. 41).
84. Canadian Parks Service (1990e, p. 31).
85. Canadian Parks Service (1991b, p. 34).
86. Canadian Parks Service (1990e, p. 31).
87. Lucas (1970, p. 303).
88. Parks Canada (1979, p. 41).
89. Canadian Parks Service (1991b, p. 34); Environment Canada (1990, p. 25); Canadian Parks Service (1990d, p. 11).
90. Canadian Parks Service (1991b, p. 35).
91. Canadian Parks Service (1991b, p. 35).
92. Parks Canada (1979, p. 43).
93. Canadian Assembly on National Parks (1986, p. 80).
94. Cited in Canadian Parks Service (1990b, p. 25).
95. Canadian Parks Service (1991b, p. 35).
96. Canadian Parks Service (1992f, p. 8).
97. Parks Canada (1979, p. 30).

Chapter 4

1. Wilson (1989, p. 97).
2. Kaufman (1960, pp. 47–48).
3. U.S. National Park Service (1988a, p. 2).
4. U.S. National Park Service (1974, pp. 17–18, 31).
5. Letter from U.S. Fish and Wildlife Service to NPS regional director dated October, 1979; reprinted in U.S. National Park Service (1988a, p. 333).
6. Letter from Townsley to Wally Steucke, 1981, as reprinted in U.S. National Park Service (1988a, p. 342).
7. Townsley letter, 1981, in U.S. National Park Service (1988a, p. 343).
8. U.S. National Park Service (1988a, p. iv).
9. U.S. National Park Service (1991d, p. 29).
10. U.S. National Park Service (1990b, p. 9; 1991d, p. 42).
11. Chase (1987b, chapters 12 and 13); Hartzog (1988, p. 254); McNamee (1984, pp. 176–77).
12. U.S. National Park Service (1988a, p. 17).
13. Bella (1987, p. 165).
14. Canadian Parks Service (1990e, p. 197).
15. Canadian Parks Service (1992d, p. 3).
16. U.S. National Park Service (1976a, p. 23).
17. Abbey in Crumbo (1981, pp. iii–iv).
18. U.S. National Park Service (1979b, p. II-20); Borden (1976, p. 74); Fradkin (1981, p. 220).
19. Thompson, Rogers, and Borden (1974, p. 28).

20. Parent and Robeson (1976, pp. iv, v).

21. Gladden (1990, pp. 87–88); *United States* v. *Brown,* 552 F.2d 821 (8th Cir. 1977).

22. *U.S. Code Annotated*, title 16, sec. 20b, 118–19.

23. U.S. National Park Service (1989a, p. G-7).

24. Address to the Conference of National Park Concessioners, March 9, 1981, as cited in Frome (1992, p. 41).

25. U.S. National Park Service (1989a, p. B-14).

26. Canadian Parks Service (1988b, p. 56).

27. Masterman (1991, p. A1).

28. Kaegi (1992, p. 14).

29. Canadian Parks Service (1988b, p. 85).

30. Muir (1912, p. 8).

31. Muir (1912, p. 7).

32. Quoted in Everhart (1983, p. 65).

33. Runte (1990, p. 7).

34. U.S. National Park Service (1980d, pp. vii, 1, 17, 32, 45, 47).

35. Cited in Conservation Foundation (1985, p. 298).

36. U.S. National Park Service (1987b, p. 4).

37. Nolte (1990c, A20).

38. U.S. Congress, Senate Subcommittee on Public Lands, National Parks, and Forests (1990, p. 115).

39. Cited in Diringer (1991, p. A3).

40. Cited in Diringer (1991, p. A3).

41. Lothian (1987, p. 24).

42. Thomas's foreword in Luxton (1975); Baird (1977, p. 89).

43. Bella (1986, p. 17).

44. Marty (1984, p. 133; 1985, p. 24). Vollmershausen quoted in Leighton (1985, p. 21).

45. Canadian Nature Federation (1980, p. 26).

46. Masterman (1989, p. B6).

47. Howse (1989, p. 48).

48. Jacobson was interviewed on June 30, 1992.

49. Canadian Parks Service (1988a, pp. 4, 16).

50. Zinkan was interviewed on July 2, 1992.

51. Leighton (1985, p. 18).

52. U.S. National Park Service (1983, p. 28).

53. U.S. National Park Service (1983, pp. 30–37).

54. Conners (1988, p. 99).

55. Diglio (1989, p. A1).

56. U.S. Congress, House Subcommittee on Environment, Energy, and Natural Resources Hearings (1990, pp. 44–47).

57. Masters (1991, p. D6).

58. Canadian Department of the Interior (1936, p. 1); Canadian Parks Service (1990e, p. 160).

59. McNamee (1992, p. 40).

60. Rowe (1990, p. 32).
61. Canadian Parks Service (1992, p. 2).

Chapter 5

1. U.S. National Park Service (1992b, p. 9).
2. Conservation Foundation (1985, p. 297).
3. National Parks and Conservation Association (1992e, p. 16); Pritchard (1992, p. 4).
4. Soden (1991, p. 571).
5. Frome (1992, p. 11).
6. Minister of the Environment's Task Force on Park Establishment (1987, pp. 4–5).
7. Landau and Eagle (1981, p. 41).
8. Statement of Representative Johnson in U.S. Congress, *Congressional Record* (1946, p. 4634).
9. Clarke and McCool (1985, p. 110).
10. Foss (1960, p. 186).
11. Culhane (1981, p. 232); Kaufman (1960, p. 236); Wilson (1989, p. 7).
12. Kaufman (1960, p. 198).
13. Tipple and Wellman (1991, p. 423).
14. Robert H. Nelson (1982, p. 66).
15. Dorfman (1993, p. 81).
16. Solow (1991, p. 187).
17. Hardin (1993, pp. 9–10).
18. Pigou (1920, p. 29).
19. Smith (1776, p. 306).
20. Clawson (1983, p. 167).
21. CQ Press (1990, pp. 693–94).
22. U.S. Congress, House Committee on Interior and Insular Affairs (1988, p. 1).
23. Gottlieb (1989, p. 9).
24. Kennedy was interviewed on the "Today" show, June 3, 1993.
25. Durant (1992, p. 238); Foresta (1984, p. 104).
26. Clarke and McCool (1985, p. 8).
27. Arnberger (1987).
28. Gordon (1989, pp. 16–17).
29. Sonnichsen (1987, p. 123); Wholey (1987, p. 183).
30. Downs and Larkey (1986, p. 246).
31. Leman (1984, p. 121).
32. Kaufman (1960, p. 161).
33. Kaufman (1960, pp. 170–75).
34. Cohen and Brand (1993, p. 14).
35. Hayek (1948, p. 79); Kelman (1987, p. 282); Lipsky (1980, p. 199).
36. Shakespeare, *Julius Caesar*, act III, scene 1.

References

Abbey, Edward. 1968. *Desert Solitaire*. New York: Ballantine Books.

———. 1981. "Foreword." In *A River Runner's Guide to the History of the Grand Canyon*, edited by Kim Crumbo, iii–iv. Boulder: Johnson Books.

Aberbach, Joel D., Robert D. Putnam, and Bert A. Rockman. 1981. *Bureaucrats and Politicians in Western Democracies*. Harvard University Press.

Aberbach, Joel D., and Bert A. Rockman. 1988. "Mandates or Mandarins? Control and Discretion in the Modern Administrative State." *Public Administration Review*, 48 (2): 606–12.

Adams, Gerald D. 1993. "The Presidio Puzzle." *Image* 3 (28): 9–15.

Adie, Robert F., and Paul G. Thomas. 1982. *Canadian Public Administration: Problematical Perspectives*. Scarborough, Ont.: Prentice-Hall Canada.

Adler, Jerry. 1986a. "The Fall of the Wild." *Newsweek*, July 28, pp. 52–54.

———. 1986b. "The Grand Illusion." *Newsweek*, July 28, pp. 48–51.

Albright, Horace M., and Robert Cahn. 1985. *The Birth of the National Park Service: The Founding Years, 1913–33*. Salt Lake City: Howe Brothers.

Allen, Glen. 1991. "Savoring Victories." *Maclean's* 104 (12): 14–15.

Allen, William. 1991. "Cruise Ships' Anchors Damaging Coral Reefs." *St. Louis Post-Dispatch*, October 22, p. 1A.

———. 1993. "Cuts, Budget Crunch Threatening Parks." *St. Louis Post-Dispatch*, May 31, pp. 1A, 10A.

Allin, Craig W. 1987. "Wilderness Preservation as a Bureaucratic Tool." In *Federal Lands Policy*, edited by Philip O. Foss, 127–38. New York: Greenwood Press.

Anderson, Jack, and Michael Binstein. 1992. "Two Faces of the Corps of Engineers." *Washington Post,* October 18, p. C7.

Anderson, Terry L., and Peter J. Hill. 1981. "Property Rights as a Common Pool Resource." In *Bureaucracy vs. Environment*, edited by John Baden and Richard L. Stroup, 22–45. University of Michigan Press.

Aniskowicz, B. T. 1990. "Life or Death?" *Nature Canada* 19 (2): 35–38.

Arnberger, Robert L. 1987. "The Politics of Community Resource Management." Presented to the Superintendency Course at the Delaware Water Gap on June 18.

Associated Press. 1992. "GAO: Sweetheart Deals in U.S. Parks." *Washington Post*, May 22, p. A4.

Aucoin, Peter. 1988. "The Mulroney Government, 1984–88: Priorities, Positional Policy and Power." In *Canada under Mulroney: An End-of-Term Report*, edited by Andrew B. Gollner and Daniel Salee, 335–56. University of Toronto Press.

Aucoin, Peter, and Herman Bakvis. 1988. *The Centralization-Decentralization Conundrum: Organization and Management in the Canadian Government*. Halifax: The Institute for Research on Public Policy.

Ayres, B. Drummond, Jr. 1988. "House Passes Bill to Enlarge Park at Battlefield in Virginia." *New York Times*, August 11, p. A18.

———. 1991. "Pollution Shrouds Shenandoah Park." *New York Times*, May 2, p. A9.

Baird, David M. 1977. *Banff National Park: How Nature Carved Its Splendor*. Edmonton: Hurtig Publishers.

Bakvis, Herman. 1989. "Regional Politics and Policy in the Mulroney Cabinet, 1984–88: Towards a Theory of the Regional Minister System in Canada." *Canadian Public Policy* 15 (2): 121–34.

Bakvis, Herman, and William M. Chandler. 1987. "Federalism and Comparative Analysis." In *Federalism and the Role of the State*, edited by Herman Bakvis and William M. Chandler, 3–11. University of Toronto Press.

Barker, Jeff. 1987. "U.S. Loses Ground on Some Park Fees." *Washington Post*, March 26, p. A25.

Barnett, Philip S. 1988. "The Mining in the Parks Act: Theory and Practice." In *Our Common Lands: Defending the National Parks*, edited by David J. Simon, 415–23. Washington, D.C.: National Parks and Conservation Association.

Barnett, Vicki. 1992. "Fragile Ecosystem Attacked From All Sides." *Calgary Herald*, February 5, p. A5.

Barney, Gerald O., P. H. Freeman, and C. A. Ulinski. 1981. *Global 2000: Implications for Canada*. Toronto: Pergamon Press.

Barr, Stephen. 1992. "Park Service Hire Is a Source of Contention." *Washington Post*, December 24, p. A11.

Barrett, Suzanne, and Don Huff. 1986. "Nielsen Reports Deliver Positive Thrusts and Potential Threats." *Seasons* 26 (2): 4–6.

Bartimus, Ted. 1991. "National Park Rangers Struggle with Low Pay, Poor Housing." *St. Louis Post-Dispatch*, July 21, p. 71G.

Bartlett, Richard A. 1985. *Yellowstone: A Wilderness Besieged*. University of Arizona Press.

Beaman, Jay, and Dick Stanley. 1991. "Counting Visitors at National Parks: Concepts and Issues." Canadian Parks Service (unpublished).

Begin, Patricia. 1989. *Use of Canada's National Parks from Resource Protection to Tourism*. Ottawa: Library of Parliament.

Bella, Leslie. 1986. "Planning in the Tourist Towns of Canada's National Parks." Working Paper 20. University of Winnipeg, Insitute of Urban Studies.

———. 1987. *Parks for Profit*. Montreal: Harvest House.

Bercuson, David, J. L. Granatstein, and W. R. Young. 1986. *Sacred Trust? Brian Mulroney and the Conservative Party in Power*. Toronto: Doubleday Canada.

Bergman, Brian. 1986. "National Parks Going Downhill?" *Equinox*, no. 29: 115–16.

Bodi, F. Lorraine. 1988. "Hydropower, Dams, and the National Parks." In *Our Common Lands: Defending the National Parks*, edited by David J. Simon, 449–63. Washington, D.C.: National Parks and Conservation Association.

Bohn, Glenn. 1985. "Logging Urged in Pacific Rim National Park." *Vancouver Sun*, August 24, p. A3.

Bohn, Glenn, and Tom Barrett. 1987. "BC Tourist Industry Calls for National Park in Charlottes." *Vancouver Sun*, June 19, p. B1.

Borden, F. Yates. 1976. "User Carrying Capacity for River-Running: The Colorado River in the Grand Canyon." National Park Service.

BOREAS. 1991. *Global Change and Biosphere Atmosphere Interaction in the Boreal Forest Biome*. National Aeronautics and Space Administration.

Borins, Sanford F. 1988. "Public Choice: 'Yes, Minister' Made It Popular, but Does Winning the Nobel Prize Make it True?" *Canadian Public Administration* 31 (1): 12–26.

Brandborg, Stewart M. 1970. "The Wilderness Law and the National Park System of the United States." In *Canadian Parks in Perspective*, edited by J. G. Nelson, 264–83. Quebec: Harvest House.

Brooks, Lloyd. 1970. "Planning a Canadian National Park System—Progress and Problems." In *Canadian Parks in Perspective*, edited by J. G. Nelson, 313–20. Quebec: Harvest House.

Brower, Kenneth. 1990. *Yosemite: An American Treasure*. Washington, D.C.: National Geographic Society.

Brown, Ben. 1991. "Federal Lands: User-Friendly." *USA Today*, June 20, p. 12C.

Brown, Robert Craig. 1970. "The Doctrine of Usefulness: Natural Resources and National Park Policy in Canada, 1887–1914." In *Canadian Parks in Perspective*, edited by J. G. Nelson, 46–62. Quebec: Harvest House.

Brue, Mike. 1989. "National Park Plan Concerns Townsfolk." *Calgary Herald*, December 23, 1989, p. F7.

Bryan, Rorke. 1973. *Much Is Taken, Much Remains: Canadian Issues in Environmental Conservation*. North Scituate, Mass.: Duxbury Press.

Bryden, Joan. 1987. "Calgary Office Stays Put in New National Park Deal." *Calgary Herald*, July 8, p. B2.

Budget of the United States Government. Various years.

Burton, T. L. 1979. "The Promises and Problems of Coordination in Parks Development and Management." In *The Canadian National Parks: Today and Tomorrow Conference II,* edited by J. G. Nelson, 313–27. University of Waterloo.

Cahn, Robert. 1993. "Science and the National Parks." *Environment* 35 (2): 25–27.

Cahn, Robert, and Patricia Cahn. 1987. "Disputed Territory." *National Parks* 61 (May/June): 28–33.

———. 1992. "Parallel Parks." *National Parks* 60 (January/February): 24–29.

Cairns, Robert D. 1992. "Natural Resources and Canadian Federalism: Decentralization, Recurrent Conflict, and Resolution." *Publius* 22 (1): 55–70.

Campbell, Colin. 1983. *Governments under Stress: Poltical Executives and Key Bureaucrats in Washington, London, and Ottawa*. University of Toronto Press.

———. 1988. "Mulroney's Broker Politics." In *Canada under Mulroney*, edited by Andrew B. Gollner and Daniel Salee, 309–44. University of Toronto Press.

Canadian Assembly on National Parks. 1986. *Heritage for Tomorrow*. Environment Canada.

Canadian Chamber of Commerce. 1990. *Organization of the Government of Canada*. Canada Communications Group.

Canadian Department of the Interior. 1936. *Prince Albert National Park*. Department of the Interior.

Canadian Environmental Advisory Council. 1991. *A Protected Areas Vision for Canada*. Minister of Supply and Services.

Canadian Government. 1990. *Canada's Green Plan*. Minister of Supply and Services.

Canadian Nature Federation. 1979. "Major Changes in Parks Policy." *Nature Canada* 8 (3): 21.

———. 1980. "Twinning the Trans-Canada Highway in Banff National Park." *Nature Canada* 9 (2): 26.

———. 1981. "Beginnings." *Nature Canada* 10 (2): 10–13.

———. 1991. "National Parks Policy Review: Paralysis through Analysis." *Nature Canada* 1 (1): 3.

———. 1992. "Caracas Declares for Parks." *Nature Alert* 2 (2): 4.

Canadian Parks Service. 1987. *Prince Albert National Park Management Plan Summary*. Environment Canada.

———. 1988a. *Banff National Park Management Plan*. Environment Canada.

———. 1988b. *Jasper National Park Management Plan*. Environment Canada.

———. 1988c. *Waskesiu Community Plan*. Environment Canada.

———. 1989a. "Impact of the Canadian Parks Service on the Economy of Alberta." Canadian Parks Service.

———. 1989b. *Private-Sector Construction in Banff, Jasper, and Waterton National Parks in Alberta*. Environment Canada.

———. 1989c. *Visitor Profile and Economic Impact Statement of Northern National Parks/ Reserves and Historic Sites: Summary Report*. Environment Canada.

———. 1990a. *Choosing Our Destiny*. Environment Canada.

———. 1990b. *Final Report of the Parks and People Task Force*. Environment Canada.

———. 1990c. *National Parks System Plan*. Environment Canada.

———. 1990d. *Report of the Science and Protection Task Force*. Environment Canada.

———. 1990e. *State of the Parks*. Environment Canada.

———. 1990f. *Strategic Plan*. Environment Canada.

———. 1991a. *Environment Canada Integration Task Force*. Environment Canada.

———. 1991b. *Proposed Policy*. Environment Canada.

———. 1991c. *Revenue Profile*. Canadian Parks Service.

———. 1992a. *A Bridge to the Future*. Environment Canada.

———. 1992b. *Application Design Report Ecosystem Management Model Project*. Environment Canada.

———. 1992c. *Bison*. Environment Canada.

———. 1992d. *Elk Island National Park Management Plan Review*. Canadian Parks Service.

———. 1992e. *The Fires of Change.* Waskesiu: Prince Albert National Park.

———. 1992f. "Prince Albert Management Plan Newsletter #1." Environment Canada.

———. 1992g. *Visitor Participation Statistics, 1991–92.* Environment Canada.

———. 1992h. *Wolf Country.* Waskesiu: Prince Albert National Park.

Canadian Parliament, House of Commons. 1983. *Debates.* April 14: 24465.

Canadian Press. 1986. "Parks Master Plan to Strike Balance, McMillan says." *Winnipeg Free Press*, February 4, p. 14.

Canadian Public Service Function. 1990. *Public Service 2000.* Canadian Public Service Function.

Carothers, Steven T., and Bryan T. Brown. 1991. *The Colorado River through Grand Canyon.* University of Arizona Press.

Carp, Teresa. 1989. "Unseen Planners Help Nature Look Her Best." *Calgary Herald*, February 25, 1989, p. D4.

Carruthers, J. A. 1979. "Planning a Canadian National Park and Related Reserve System." In *The Canadian National Parks: Today and Tomorrow Conference II,* edited by J. G. Nelson, 645–70. University of Waterloo.

Caulfield, Henry P. 1989. "The Conservation and Environmental Movements." In *Environmental Politics and Policy: Theories and Evidence,* edited by James P. Lester, 13–56. Duke University Press.

Chase, Alston. 1987a. "How to Save Our National Parks." *The Atlantic Monthly* 263 (July): 35–44.

———. 1987b. *Playing God in Yellowstone: The Destruction of America's First National Park.* Harcourt Brace Jovanovich.

———. 1991. "Unhappy Birthday." *Outside*, December, pp. 33–40.

Chen, Ingfei. 1992. "U.S. Park Plan for Presidio Moves Ahead." *San Francisco Chronicle,* September 2, 1992, p. 7.

Chodos, Robert, Rae Murphy, and Eric Hamovitch. 1988. *Selling Out: Four Years of the Mulroney Government.* Toronto: James Lorimer and Company.

Church, Lisa. 1988a. "Controlled National Park Fires Urged." *Calgary Herald*, October 16, p. A12.

———. 1988b. "Park Hopes to Let Buffalo Roam." *Calgary Herald*, May 6, 1988, p. A18.

CIS Annual. Various years. Washington, D.C.: Congressional Information Service.

Claiborne, William. 1990. "Canada Unveils Major Environmental Plan." *Washington Post*, December 12, 1990, p. A19.

Clarke, G. B. 1979. "Tourism in National Parks." In *The Canadian National Parks: Today and Tomorrow Conference II,* edited by J. G. Nelson, 45–48. University of Waterloo.

Clarke, Jeanne Nienaber, and Daniel McCool. 1985. *Staking Out the Terrain: Power Differentials among Natural Resource Management Agencies.* State University of New York Press.

Clarkson, Stephen. 1982. *Canada and the Reagan Challenge.* Toronto: James Lorimer.

Clarkson, Stephen, and Christina McCall. 1990. *Trudeau and Our Times: The Magnificent Obsession.* Toronto: McClelland and Stewart.

Clawson, Marion. 1983. *The Federal Lands Revisited*. Washington, D.C.: Resources for the Future.

Coates, James. 1991a. "Creature Comforts Taking Toll on Park Wilderness." *Chicago Tribune*, April 21, section 1, pp. 1, 10.

———. 1991b. "Crowds Pose Threat to U.S. Park System." *Chicago Tribune*, April 21, section 1, pp. 1, 16.

Cockburn, Alexander. 1993. "Bruce Babbitt, Compromised by Compromise." *Washington Post National Weekly Edition*, September 6–12, p. 24.

Cohen, Steven, and Ronald Brand. 1993. *Total Quality Management in Government: A Practical Guide for the Real World*. San Francisco: Jossey-Boss.

Cohn, D'Vera. 1990a. "Smoggy Days Obscuring Views at Shenandoah Park." *Washington Post*, August 21, 1990, p. B1.

———. 1990b. "Va. Rejects Park Officials' Power Plant Objections." *Washington Post*, November 7, 1990, p. C5.

———. 1992. "Plan to Close Skyline Drive Stalls." *Washington Post*, November 25, 1992, p. B1.

Cohn, Roger, and Ted Williams. 1993. "Interior Views." *Audubon* 95 (3): 78–84.

Cole, Taylor. 1966. "The Canadian Bureaucracy and Federalism, 1947–1965." Manuscript from University of Denver Social Science Foundation.

Conners, John A. 1988. *Shenandoah National Park: An Interpretive Guide*. Blacksburg, Va: McDonald & Woodward.

Conservation Foundation. 1985. *National Parks for a New Generation: Visions, Realities, Prospects*. Washington, D.C.: The Conservation Foundation.

Cook, Earl. 1982. "The Role of History in Acceptance of Nuclear Power." *Social Sciences Quarterly* 63 (1): 3–15.

Cordell, H. Ken. 1990. "Outdoor Recreation and Wilderness." In *Natural Resources for the 21st Century*, edited by R. Neil Simpson and Dwight Hair, 242–68. Washington, D.C.: Island Press.

Cox, Kevin. 1987. "Sewage Problems Tainting Charms of National Park." *Toronto Globe and Mail*, September 26, p. A4.

CQ Press. 1979. *1978 CQ Almanac*. Washington, D.C.: CQ Press.

———. 1990. *1989 CQ Almanac*. Washington, D.C.: CQ Press.

Cragg, J. B. 1970. "Research in National and Provincial Parks: Possibilities and Limitations." In *Canadian Parks in Perspective*, edited by J. G. Nelson, 111–22. Quebec: Harvest House.

Craig, Bruce. 1991a. "Diamonds and Rust." *National Parks* 65 (May/June): 40–44.

———. 1991b. "Promise to Keep." *National Parks* 65 (November/December): 18–19.

Craighead, John J., and Frank C. Craighead. 1971. *Grizzly Bear-Man Relationships in Yellowstone National Park*. Washington, D.C.: National Park Service.

Crosson, Judith. 1991. "National Parks May Seek Sponsors." *San Jose Mercury News*, September 21, p. 7A.

Crozier, Michel. 1964. *The Bureaucratic Phenomenon*. University of Chicago Press.

Crumbo, Kim. 1981. *A River-Runner's Guide to the History of the Grand Canyon*. Boulder: Johnson Books.

Culhane, Paul J. 1978. "Natural Resources Policy: Procedural Change and Substantive Environmentalism." In *Nationalizing Government: Public Policies in*

America, edited by Theodore J. Lowi and Alan Stone, 201–62. Beverly Hills: Sage Publications.

———. 1981. *Public Lands Politics: Interest Group Influence on the Forest Service and the Bureau of Land Management*. Johns Hopkins.

Davidson, A. T. 1979. "Canada's National Parks: Past and Future." In *The Canadian National Parks: Today and Tomorrow Conference II*, edited by J. G. Nelson, 23–37. University of Waterloo.

Davis, Charles, and Sandra Davis. 1987. "Wilderness Protection at What Price? An Empirical Assessment." In *Federal Lands Policy*, edited by Philip O. Foss 113–25. New York: Greenwood Press.

Davis, Sandra. 1989. "Development or Conservation: A Round Table on National Parks Policy." *Canadian Parliamentary Review* 12 (4): 26–29.

Dawson, Chris. 1990. "Poacher Gets Undeserved Break." *Calgary Herald*, December 30, p. B6.

Dearden, Philip. 1985. "Too Many People Spoil the Parks." *Canadian Geographic* 105 (1): 88–89.

Deford, Susan. 1988. "Preservationists Hope to Protect Antietam from March of Development." *Washington Post*, August 18, p. Md1.

Devoto, Bernard. 1953. "Let's Close the National Parks." *Harper's Magazine* 207 (October): 49–52.

Diglio, Alice. 1989. "Development Paints Grim Vista at Shenandoah." *Washington Post*, October 7, p. A1.

Diringer, Elliot. 1989. "New Drive for Hotels in Yosemite." *San Francisco Chronicle*, December 5, p. A1.

———. 1991. "MCA Sale May Not Change Yosemite." *San Francisco Chronicle*, January 10, p. A3.

Dodd, Elizabeth M. 1988. "The Geothermal Steam Act: Unlocking Its Protective Provisions." In *Our Common Lands: Defending the National Parks*, edited by David J. Simon, 425–47. Washington, D.C.: National Parks and Conservation Association.

Doern, G. Bruce, and Richard W. Phidd. 1983. *Canadian Public Policy: Ideas, Structure, Process*. Toronto: Methuen.

Doerr, John E. 1938. *The National Parks of Canada*. Ottawa: National Parks Bureau.

Dolan, Maura. 1991. "Zion Park, Town Clash over Development." *Los Angeles Times*, May 13, pp. A1, A22, A23.

Dorfman, Robert. 1993. "Some Concepts from Welfare Economics." In *Economics of the Environment: Selected Readings*, edited by Robert Dorfman and Nancy Dorfman, 79–96. Norton.

Downs, Anthony. 1967. *Inside Bureaucracy*. Boston: Little, Brown and Company.

Downs, George W., and Patrick D. Larkey. 1986. *The Search for Government Efficiency: From Hubris to Helplessness*. Philadelphia: Temple Press.

Dunlap, Riley E., George H. Gallup, Jr., and Alec Gallup. 1992. "Worldwide Environmental Poll." *Gallup Poll Monthly* 320 (May): 42–47.

Durant, Robert F. 1992. *The Administrative Presidency Revisited: Public Lands, the BLM, and the Reagan Revolution*. Albany: State University of New York Press.

Edelson, Nathan. 1993. "Parks without People." *Wall Street Journal*, May 26, p. A18.

Editors of the *Calgary Herald*. 1984. "Parks Battle Looms." *Calgary Herald*, May 4, p. A4.

Editors of *The Economist*. 1993. "Managing Paradise." *The Economist*, February 6, p. 31.

Editors of *USA Today*. 1992. "News." *USA Today*, March 4, p. A6.

Editors of the *Washington Post*. 1987. "Parks and Preservation." *Washington Post*, August 21, p. A22.

Egan, Timothy. 1990. "Forest Service Abusing Role, Dissidents Say." *New York Times*, March 4, pp. A1, 26.

Eidsvik, Harold K., and William D. Henwood. 1990. "Canada." In *International Handbook of National Parks and Nature Reserves*, edited by Craig W. Allin, 61–81. New York: Greenwood Press.

Encyclopedia of Associations. Various. Detroit: Gale Research.

Environment Canada. 1987. *1986–87 Estimates*. Environment Canada.

———. 1988. *Legislation Governing the National Parks of Canada*. Environment Canada.

———. 1990. *Canada's Green Plan: Summary*. Environment Canada.

———. 1991. *Report on Transition—Year One*. Environment Canada.

———. 1992. *1992–93 Estimates*. Environment Canada.

Epstein, Edward. 1993. "Gorbachev to Visit S.F. Next Week." *San Francisco Chronicle*, April 7, p. A1.

Everhart, William C. 1983. *The National Park Service*. Boulder, Colorado: Westview Press.

Fabry, Judith K. 1988. *Guadalupe Mountains Administrative History*. Santa Fe: Southwest Cultural Resources Center.

Fairfax, Sally K., and Carolyn E. Yale. 1987. *Federal Lands: A Guide to Planning, Management and State Revenues*. Washington, D.C.: Island Press.

Feldman, Elliot J., and Michael A. Goldberg. 1987. "General Lessons from Diverse Cases." In *Land Rites and Wrongs: The Management, Regulation and Use of Land in Canada and the United States*, edited by Elliot Feldman and Michael Goldberg, 271–82. Cambridge: Lincoln Institute of Land Policy.

Ferejohn, John A. 1974. *Pork Barrel Politics: Rivers and Harbors Legislation*. Stanford University Press.

Ferguson, Gary. 1993. *Walking down the Wild: A Journey through the Yellowstone Rockies*. New York: Simon & Schuster.

Fisher, Louis. 1991. "Congress as Micromanager of the Executive Branch." In *The Managerial Presidency*, edited by James P. Pfiffner, 225–37. Pacific Grove, California: Brooks/Cole.

Fitzgerald, F. Scott. 1925. *The Great Gatsby*. 1986 edition, p. 182. New York: Collier Books.

Foresta, Ronald A. 1984. *America's National Parks and Their Keepers*. Washington, D.C.: Resources for the Future.

Foss, Phillip O. 1960. *Politics and Grass: The Administration of Grazing on the Public Domain*. University of Washington Press.

Fradkin, Philip L. 1981. *A River No More: The Colorado River and the West*. New York: Alfred A. Knopf.

Francis, John G., and Richard Ganzel. 1984. "Conclusion: Public Lands, Natural Resources, and the Shaping of American Federalism." In *Western Public Lands: The Management of Natural Resources in a Time of Declining Federalism,* edited by John G. Francis and Richard Ganzel, 303–05. Totowa, N. J.: Rowman and Allenheld.

Franks, C. E. S. 1987. *The Parliament of Canada.* University of Toronto Press.

Fraser, J. Keith. 1985. "A Cheer for Our Parks, a Prayer for the Rest." *Canadian Geographic,* 105 (1): 6.

Fraser, R. 1979. "Setting the Stage Discussion." In *The Canadian National Parks: Today and Tomorrow Conference II,* edited by J. G. Nelson, 54–61. University of Waterloo.

Freemuth, John. 1989. "The National Parks: Political versus Professional Determinants of Policy." *Public Administration Review* 49 (3): 278–86.

———. 1991. *Islands under Siege: National Parks and the Politics of External Threats.* University of Kansas.

Frome, Michael. 1971. *The Forest Service.* New York: Praeger Publishers.

———. 1992. *Regreening the National Parks.* University of Arizona Press.

Frome, Michael, Roland H. Waver, and Paul C. Pritchard. 1990. "United States: National Parks." In *International Handbook of National Parks and Nature Reserves,* edited by Craig W. Allin, 415–30. New York: Greenwood Press.

Fuller, W. A. 1970. "National Parks and Nature Preservation." In *Canadian Parks in Perspective,* edited by J. G. Nelson, 264–83. Quebec: Harvest House.

Fultz, Keith O. 1992. *National Park Service: Policies and Practices for Determining Concessioners' Building Use Fees.* May 21. GAO/T-RCED–92–66. General Accounting Office.

Galbraith, Paul. 1989. "An Evaluation of the Status of Aquatic Resource Management in Canadian National Parks." In *Use and Management of Aquatic Resources in Canada's National Parks,* edited by D. C. Harvey, S. J. Woodley, and A. R. Haworth, 39–52. University of Waterloo: Heritage Resources Center.

Garrett, W. E. 1978. "Are We Loving It to Death?" *National Geographic* 154 (1): 16–51.

Gates, Paul W. 1968. *History of Public Land Law Development.* Government Printing Office.

Geddes, Ashley. 1989. "Clamp Sought on Rules in Parks." *Calgary Herald,* February 9, pp. A1–A2.

Geist, V. 1979. "Critique." In *The Canadian National Parks: Today and Tomorrow Conference II,* edited by J. G. Nelson, 51–53. University of Waterloo.

Ghiglieri, Michael P. 1992. *Canyon.* University of Arizona Press.

Gibbins, Roger. 1982. *Regionalism, Territorial Politics in Canada and the United States.* Toronto: Butterworths.

Gilmour, John B. 1990. *Reconcilable Differences? Congress, the Budget Process, and the Deficit.* University of California Press.

Gladden, James N. 1990. *The Boundary Waters Canoe Area: Wilderness Values and Motorized Recreation.* Iowa State University Press.

Glick, Daniel. 1992. "Grizzlies Come Back." *Newsweek,* September 7, pp. 60–61.

Gordon, John C. 1984. "Educating Tomorrow's Foresters." *American Forests* 90 (10): 56, 61–63.

———. 1989. "The Scientific Method: NPCA's Advisory Commission Finds Park Research Neglected." *National Parks* 63 (May/June): 16–17.

Gottlieb, Alan M. 1989. *The Wise Use Agenda: The Citizen's Policy Guide to Environmental Resource Issues.* Bellevue, Wash.: The Free Enterprise Press.

Gouldner, Alvin W. 1954. *Patterns of Industrial Bureaucracy.* The Free Press.

Graham, R., P. W. Nilsen, and R. J. Payne. 1987. "Visitor Activity Planning and Management in Canadian National Parks." In *Social Science in Natural Resource Management Systems,* edited by Marc L. Miller, Richard P. Gale, and Perry J. Brown, 149–66. Boulder: Westview Press.

Gray, Brian F. 1988. "No Holier Temples: Protecting the National Parks Through Wild and Scenic River Designation." In *Our Common Lands: Defending the National Parks,* edited by David J. Simon, 331–87. Washington, D.C.: National Parks and Conservation Association.

Gwyn, Richard. 1980. *The Northern Magus: Pierre Trudeau and the Canadians.* Toronto: McLelland and Stewart.

Hall, Charles W. 1992. "Plan to Shut Part of Skyline Drive is Scuttled." *Washington Post,* December 10, p. D3.

Halperin, Morton H. 1974. *Bureaucratic Politics and Foreign Policy.* Brookings.

Harasymuk, David. 1991. "Big Trouble in Little Paradise." *Parks & Wilderness* (Winter): 14–16.

Hardin, Garrett. 1968. "The Tragedy of the Commons." In *Economics of the Environment,* edited by Robert Dorfman and Nancy Dorfman, 3d ed., 1993, 5–19. Norton.

Harris, John F. 1991. "In Shenandoah Park, an Outspoken Shepherd." *Washington Post,* November 16, p. A1.

Harrison, Kathryn, and George Hoberg. 1991. "Setting the Environmental Agenda in Canada and the United States: The Cases of Dioxin and Radon." *Canadian Journal of Political Science* 24 (1): 3–27.

Hartzog, George B., Jr. 1988. *Battling for the National Parks.* Mt. Kisco: Moyer Bell Limited.

Harvey, D. Michael. 1988. "The Federal Land Policy and Management Act: The Bureau of Land Management's Role in Park Protection." In *Our Common Lands: Defending the National Parks,* edited by David J. Simon, 127–41. Washington, D.C.: National Parks and Conservation Association.

Hatfield, Larry D. 1992. "Tour Buses Failing to Make Grade." *San Francisco Examiner,* August 16, p. B1.

Hayek, Friedrich A. von. 1948. *Individualism and Economic Order.* University of Chicago Press.

Hays, Samuel P. 1959. *Conservation and the Gospel of Efficiency.* Harvard University Press.

———. 1987. *Beauty, Health, and Permanence.* Cambridge University Press.

Heacox, Kim. 1991. "The Taming of Denali." *National Parks* 65 (February/March): 20–25.

Healey, Jon. 1992. "Jetties' Approval Politically Motivated, Opponents Say." *Winston-Salem Journal*, October 31, pp. 21, 23.

Heclo, Hugh. 1975. "OMB and the Presidency—The Problem of Neutral Competence." *Public Interest* 38 (Winter): 80–98.

———. 1977. *A Government of Strangers*. Brookings.

Herman, Dennis J. 1992. "Loving Them to Death: Legal Controls on the Type and Scale of Development in the National Parks." *Stanford Environmental Law Journal* 11: 3–67.

Hill, Larry. 1985. "New Minister Opposes Mining in National Parks." *Winnipeg Free Press*, August 21, p. 8.

Hinds, Michael deCourcy. 1992. "Much Steaming over 'Steamtown.'" *New York Times*, February 8, p. 12.

Hoberg, George. Forthcoming. "Governing the Commons: Environmental Policy in Canada and the United States." In *Canada and the United States in a Changing World*, edited by Richard Simeon and Keith Banting.

Hocker, Philip. 1988. "Oil, Gas, and Parks." In *Our Common Lands: Defending the National Parks*, edited by David J. Simon, 389–413. Washington, D.C.: National Parks and Conservation Association.

Howse, John. 1989. "An Identity Crisis." *MacLean's*, February 27, p. 48.

Hoy, Claire. 1987. *Friends in High Places: Politics and Patronage in the Mulroney Government*. Toronto: Key Porter Books.

Huber, Jeanne. 1992a. "Officials Get Angry Earful at Yosemite Hearing." *San Jose Mercury News*, February 2, p. 3B.

———. 1992b. "Yosemite's 'Death Sentence'." *San Jose Mercury News*, January 29, p. 1A

Huff, Don. 1985. "Good News Comes out of Parks Conference." *Seasons* 25 (4): 8.

Hume, Mark. 1988. "Group Wants a National Park for Fraser River Delta's Birdlife." *Vancouver Sun*, February 6, p. H18.

Hume, Stephen. 1990. "CanFor Shows What It Thinks of a National Park." *Vancouver Sun*, December 19, p, A13.

Hummel, Don. 1987. *Stealing the National Parks*. Bellevue, Wash.: The Free Enterprise Press.

Hummel, Monte. 1985. "Environmental and Conservation Movements." In *The Canadian Encyclopedia*, edited by James H. Marsh, 585–86. Edmonton: Hurtig Publishers.

Hunt, Constance D. 1990. "A Note on Environmental Impact Assessment in Canada." *Environmental Law Journal* 20: 789–810.

Ingram, Helen M., and Dean E. Mann. 1989. "Interest Groups and Environmental Policy." In *Environmental Politics and Policy: Theories and Evidence*, edited by James P. Lester, 135–57. Duke University Press.

Interagency Grizzly Bear Study Team. 1975. *Yellowstone Grizzly Bear Investigations: 1974 Annual Report*. Washington, D.C.: National Park Service.

———. 1991. *Yellowstone Grizzly Bear Investigations: 1990 Annual Report*. Washington, D.C.: National Park Service.

Israelson, David. 1989. "National Parks to Grow Four-Fold, Minister Vows." *Toronto Star*, December 14, p. A2.

Johnson, M. Bruce. 1981. "The Environmental Costs of Bureaucratic Governance: Theory and Cases." In *Bureaucracy vs. Environment*, edited by John Baden and Richard L. Stroup, 217–23. University of Michigan Press.

Jones, William R. 1989. *Yosemite: The Story behind the Scenery*. Kansas City: KC Publications.

Junkin, Elizabeth Darby. 1986. *Lands of Brighter Destiny*. Golden, Colo.: Fulcrum.

Kaegi, Dave. 1992. "A Vision for the Maligne." *Participation* (Spring): 14.

Kamen, Al. 1993. "Clinton Selects a Jurist, Sort of." *Washington Post*, May 24, p. A17.

Kanamine, Linda. 1993. "Budget Crisis Poses Threat to Treasures." *USA Today*, February 3, pp. 1A, 6A.

Kaplan, David A. 1991. "Born Free, Sold Dear." *Newsweek*, May 6, pp. 52–54.

Katzmann, Robert A. 1981. *Regulatory Bureaucracy: The Federal Trade Commission and Antitrust Policy*. MIT Press.

Kaufman, Herbert. 1960. *The Forest Ranger*. Johns Hopkins Press.

Keiter, Robert B. 1988. "National Park Protection: Putting the Organic Act to Work." In *Our Common Lands: Defending the National Parks,* edited by David J. Simon, 75–83. Washington, D.C.: National Parks and Conservation Association.

Kelman, Steven. 1981. "Cost-Benefit Analysis: An Ethical Critique." *Regulation* 12 (January/February): 33–40.

———. 1987. *Making Public Policy*. Basic Books.

Kenski, Henry C., and Margaret Corgan Kenski. 1984. "Congress against the President." In *Environmental Policy in the 1980s*, edited by Norman J. Vig and Michael E. Kraft, 97–120. Washington, D.C.: CQ Press.

Kenworthy, Tom. 1991a. "Manipulation Charged on Yellowstone Report." *Washington Post*, November 1, p. A4.

———. 1991b. "No 'Lasting Damage' Seen from Dude Ranch in Park." *Washington Post*, September 22, p. A4.

———. 1991c. "Park Service Dissent Omitted in Report on Yellowstone." *Washington Post*, October 20, p. A3.

———. 1992. "Stable Project Rides Again." *Washington Post*, December 11, p. A25.

Kilgour, David, and John Kirsner. 1989. "Discipline versus Democracy." *Parliamentary Government* 8 (1): 21–22.

Knott, Jack, and Gary Miller. 1987. *Reforming Bureaucracy: The Politics of Institutional Choice*. Prentice-Hall.

Koenig, Robert L. 1991. "2 Bills Seek to Push Arch Park Extension." *St. Louis Post-Dispatch*, August 4, p. 12.

Kraft, Michael E., and Norman J. Vig. 1990. "Environmental Policy from the Seventies to the Nineties: Continuity and Change." In *Environmental Policy in the 1990s,* edited by Norman J. Vig and Michael E. Kraft, 3–31. Washington, D.C.: CQ Press.

Kremp, Sabine. 1981. "A Perspective on BLM Grazing Policy." In *Bureaucracy vs. Environment*, edited by John Baden and Richard L. Stroup, 124–53. University of Michigan Press.

Krutilla, John V. 1967. "Conservation Reconsidered." In *Economics of the Environment*, edited by Robert Dorfman and Nancy Dorfman, 3d ed., 1993, 188–98. Norton.

Lamb, Mike. 1991. "Parks Service Trims Staffing Levels." *Calgary Herald*, April 1, p. B2.

Lancaster, John. 1990a. "Parks, Perks and Pork." *Washington Post*, December 1, p. A1.

————. 1990b. "Profiteering in the Parks?" *Washington Post*, May 25, p. A19.

————. 1991. "Japanese Firm Agrees to Sell Park Business." *Washington Post*, January 19, p. A1.

Landau, Martin, and Eva Eagle. 1981. "On the Concept of Decentralization." Berkeley: University of California Project on Managing Decentralization, manuscript.

Lash, Jonathan. 1984. *A Season of Spoils*. Pantheon Books.

Leeson, B. F. 1979. "Changing Use Patterns in Canada's Western National Parks." In *The Canadian National Parks: Today and Tomorrow Conference II*, edited by J. G. Nelson, 83–86. University of Waterloo.

Leighton, Douglas. 1985. "Banff Today: Struggling to Cope with Success." *Canadian Geographic* 105 (1): 16–21.

Leiss, William. 1979. "Political Aspects of Environmental Issues." In *Ecology versus Politics in Canada*, edited by William Leiss. University of Toronto Press.

Leman, Christopher K. 1984. "How the Privatization Revolution Failed, and Why Public Land Management Needs Reform Anyway." In *Western Public Lands: The Management of Natural Resources in a Time of Declining Federalism*, edited by John G. Francis and Richard Ganzel, 110–28. Totowa, N.J.: Rowman and Allenheld.

————. 1987a. "The Concepts of Public and Private and their Applicability to North American Lands." In *Land Rites and Wrongs*, edited by Elliot Feldman and Michael Goldberg, 23–37. Cambridge: Lincoln Institute of Land Policy.

————. 1987b. "A Forest of Institutions: Patterns of Choice on North American Timberlands." In *Land Rites and Wrongs*, edited by Elliot Feldman and Michael Goldberg, 149–200. Cambridge: Lincoln Institute of Land Policy.

Lemons, John, and Dean Stout. 1984. "A Reinterpretation of National Park Legislation." *Environmental Law Journal* 15: 41–65.

Lester, James P. 1990. "A New Federalism: Environmental Policy in the States." In *Environmental Policy in the 1990s*, edited by Norman J. Vig and Michael E. Kraft, 59–79. Washington, D.C.: CQ Press.

Lever, Nora S. 1988. "What's Happened under the New Rules?" *Canadian Parliamentary Review* (Autumn): 14–16.

Lewis, J. E. 1979. "Federal-Provincial Relations." In *The Canadian National Parks: Today and Tomorrow Conference II*, edited by J. G. Nelson, 69–72. University of Waterloo.

Libecap, Gary D. 1981a. "Bureaucratic Opposition to the Assignment of Property Rights: Overgrazing on the Western Range." *Journal of Economic History* 41 (1): 151–58.

————. 1981b. *Locking up the Range*. Cambridge, Mass.: Ballinger.

Linden, Eugene. 1992. "Search for the Wolf." *Time*, November 9, p. 22.

Lipset, Seymour Martin. 1990. *Continental Divide*. New York: Routledge.

Lipsky, Michael. 1980. *Street-Level Bureaucracy*. New York: Russell Sage Foundation.

Lockhart, William J. 1988. "External Park Threats and Interior's Limits: The Need for an Independent Park Service." In *Our Common Lands: Defending the National Parks*, edited by David J. Simon, 3–72. Washington, D.C.: National Parks and Conservation Association.

Lothian, W. F. 1987. *A Brief History of Canada's National Parks*. Environment Canada.

Lowey, Mark. 1985. "National Park 'Jewels' Are Losing Their Shine." *Calgary Herald*, July 13, p. A5.

———. 1987. "$75m Urged for New National Parks." *Calgary Herald*, June 1, pp. A1, A2.

Lowi, Theodore J. 1964. "American Business, Public Policy, Case-Studies, and Political Theory." *World Politics* 16 (July): 677–715.

———. 1979. *The End of Liberalism*. 2d ed. Norton.

Lowry, William R. 1993. "Land of the Fee: Entrance Fees and the National Park Service." *Political Research Quarterly* 46 (December): 823–45.

Lubin, Martin. 1986. "Public Policy, Canada and the U.S." *Policy Studies Journal* 14 (4): 555–65.

Lucas, Robert C. 1970. "Research Needed for National Parks." In *Canadian Parks in Perspective*, edited by J. G. Nelson, 284–309. Quebec: Harvest House.

Luxton, Eleanor G. 1975. *Banff, Canada's First National Park: A History and Memory of Rocky Mountains Park*. Banff: Summerthought.

Maass, Arthur. 1951. *Muddy Waters: The Army Engineers and the Nation's Rivers*. Harvard University Press.

Mackintosh, Barry. 1984. *The National Parks: Shaping the System*. Washington, D.C.: Department of the Interior.

Magnuson, Warren. 1990. "Critical Social Movements: De-centring the State." In *Canadian Politics: An Introduction to the Discipline*, edited by Alain-g. Gagnon and James P. Bickeston. Peterborough, Ontario: Broadview Press.

Magraw, Daniel Barstow. 1988. "International Law and Park Protection: A Global Responsibility." In *Our Common Lands: Defending the National Parks*, edited by David J. Simon, 143–73. Washington, D.C.: National Parks and Conservation Association.

Malcolm, Andrew H. 1989. "Visitors Stream back to Yellowstone after the Fires of 1988." *New York Times*, September 3, p. A30.

Marsh, John S. 1970. "Maintaining the Wilderness Experience in Canada's National Parks." In *Canadian Parks in Perspective*, edited by J. G. Nelson, 123–36. Quebec: Harvest House.

———. 1979. "Recreation and the Wilderness Experience in Canada's National Parks." In *The Canadian National Parks: Today and Tomorrow Conference II*, edited by J. G. Nelson, 151–77. University of Waterloo.

———. 1982. "Concluding Remarks." In *Parks and Tourism: Progress or Prostitution?* edited by Bruce Downie and Bob Peart, 64–66. Victoria: National and Provincial Parks of Canada.

Marty, Sid. 1984. *A Grand and Fabulous Notion: The First Century of Canada's Parks*. Toronto: NC Press Limited.

————. 1985. "Please Don't Coddle the Wild Animals." *Canadian Geographic* 105 (1): 22–29.

Masterman, Bruce. 1989. "Values Have Changed in National Parks." *Calgary Herald,* May 21, p. B6.

————. 1991. "DNA Enters War against Poaching." *Calgary Herald,* October 27, p. A1.

Masters, Brooke A. 1991. "Power Plants Harmless to Shenandoah, EPA Says." *Washington Post,* January 31, p. D6.

Mazmanian, Daniel A., and Jeanne Nienaber. 1979. *Can Organizations Change?* Brookings.

McAllister, Bill. 1991a. "Seeking More Competition in the Parks." *Washington Post,* August 29, p. A21.

————. 1991b. "Wrangle over Grand Teton Ranch." *Washington Post,* August 29, p. A21.

McCombs, Phil. 1985. "Holiday in the Park with Bill." *Washington Post,* June 28, pp. D1-D2.

McConnell, Grant. 1966. *Private Power and American Democracy.* Houghton Mifflin.

McGinley, Patrick. 1988. "The Surface Mining Control and Reclamation Act: Ten Years of Promise and Problems for the National Parks." In *Our Common Lands: Defending the National Parks,* edited by David J. Simon, 465–98. Washington, D.C.: National Parks and Conservation Association.

McLaren, Christie. 1985. "Saving the Wilderness." *Toronto Globe and Mail,* September 3, pp. 1, 16.

————. 1986. "Creation of More National Parks Urged." *Toronto Globe and Mail,* June 2, p. A10.

McManus, Reed. 1992. "Ambushed from Within." *Sierra* 77 (May): 43–45.

McNamee, Kevin. 1992. "Out of Bounds." *Nature Canada* 21 (1): 34–41.

McNamee, Thomas. 1984. *The Grizzly Bear.* Alfred A. Knopf.

Meier, Barry. 1993. "With Rescue Costs Growing, U.S. Considers Billing the Rescued." *New York Times,* March 28, p. A12.

Melnick, R. Shep. 1983. *Regulation and the Courts: The Case of the Clean Air Act.* Brookings.

Minister of the Environment's Task Force on Park Establishment. 1987. *Our Parks— Vision for the 21st Century.* Environment Canada.

Mintzmyer, Lorraine. 1992. "Disservice to the Parks." *National Parks* 66 (November/December): 24–25.

Mitchell, Brent. 1991. "Park Service Problems." *Washington Post,* August 21, p. A19.

Mitchell, Robert Cameron. 1990. "Public Opinion and the Green Lobby: Poised for the 1990s?" In *Environmental Policy in the 1980s,* edited by Norman J. Vig and Michael E. Kraft, 81–99. Washington, D.C.: CQ Press.

Moe, Terry M. 1985. "The Politicized Presidency." In *The New Direction in American Politics,* edited by John Chubb and Paul E. Peterson, 235–71. Brookings.

————. 1989. "The Politics of Bureaucratic Structure." In *Can the Government Govern?* edited by John Chubb and Paul E. Peterson, 267–329. Brookings.

Monastersky, Richard. 1988. "Lessons from the Flames." *Science News,* November 12, pp. 314–17.

Muir, John. 1912. *The Yosemite.* 1986 ed. University of Wisconsin Press.

Munro, Margaret. 1987. "National Parks: Greed Triumphs over Pride." *Montreal Gazette,* June 22, p. A7.

Nash, Roderick. 1982. *Wilderness and the American Mind.* 3d ed. Yale University Press.

Nathan, Richard P. 1983. *The Administrative Presidency.* New York: Wiley.

National Park Service Organic Act. U.S. Code Annotated. Title 16, sec. 1.

National Parks and Conservation Association. 1988. *Investing in Park Futures.* Washington, D.C.: NPCA.

———. 1991a. "NPCA News." *National Parks* 65 (November/December): 8–16.

———. 1991b. "NPS Personnel Moves Concern Observers." *National Parks* 65 (May/June): 18–19.

———. 1992a. "Geothermal Drilling Sparks Hot Debate." *National Parks* 66 (January/February): 8–9.

———. 1992b. "New Evidence out in Yellowstone Probe." *National Parks* 66 (August/September): 8–9.

———. 1992c. "1992 Park Service Budget Released." *National Parks,* 66 (January/February): 10.

———. 1992d. " 'Pork' Projects Drain Park Service Funds." *National Parks* 66 (May/June): 12.

———. 1992e. *Parks in Peril: The Race against Time Continues.* Washington, D.C.: National Parks and Conservation Association.

———. 1992f. "Study Finds Overhaul of Park Science Needed." *National Parks* 66 (November/December): 16.

———. 1993a. "Babbitt Puts New Focus on Parks." *National Parks* 67 (May/June): 10–11.

———. 1993b. "Parks Hit Hard by Budget Cuts." *National Parks* 67 (March/April): 8–9.

———. 1993c. "Yosemite Contract Is Step toward Reform." *National Parks* 67 (May/June): 13.

Nelson, Barry, and Bruce Patterson. 1979. "Park Responsibility Shifted." *Calgary Herald,* June 7, p. A1.

Nelson, J. G. 1970a. "Introduction." In *Canadian Parks in Perspective,* edited by J. G. Nelson, 9–15. Quebec: Harvest House.

———. 1970b. "Man and Landscape Change in Banff National Park: A National Park Problem in Perspective." In *Canadian Parks in Perspective,* edited by J. G. Nelson, 63–96. Quebec: Harvest House.

———. 1979. "An Introduction." In *The Canadian National Parks: Today and Tomorrow Conference II,* edited by J. G. Nelson, 15–21. University of Waterloo.

Nelson, Robert H. 1982. "The Public Lands." In *Current Issues in Natural Resource Policy,* edited by Paul R. Portney, 14–73. Washington, D.C.: Resources for the Future.

Nemetz, Peter N., W. T. Stanbury, and Fred Thompson. 1986. "Social Regulation in Canada: An Overview and Comparison with the American Model." *Policy Studies Journal* 14 (4): 580–603.

Newman, Anne. 1993. "Volunteers Help Parks Weather Cutbacks." *Wall Street Journal*, May 19, p. B1.

New York Times News Service. 1993. "Mounting Cost of Park Rescues Sparks Debate." *St. Louis Post-Dispatch*, March 29, p. 9A.

Nicol, J. I. 1970. "The National Parks Movement in Canada." In *Canadian Parks in Perspective*, edited by J. G. Nelson, 19–34. Quebec: Harvest House.

Nienaber, Jeanne, and Aaron Wildavsky. 1973. *The Budgeting and Evaluation of Federal Recreation Programs.* Basic Books.

Nikiforuk, Andrew. 1990. "Islands of Extinction." *Equinox* 52 (July/August): 30–43.

Noll, Roger G. 1985. "Government Regulatory Behavior." In *Regulatory Policy and the Social Sciences*, edited by Roger G. Noll, 9–63. University of California Press.

Nolte, Carl. 1990a. "Bargain Sale Proposed for Yosemite Concession." *San Francisco Chronicle*, December 27, p. A1.

———. 1990b. "Yosemite Franchise Provokes New Outcry." *San Francisco Chronicle*, April 4, p.A3.

———. 1990c. "Yosemite Is 100-and Ailing." *San Francisco Chronicle*, September 28, p. A1.

Nudds, Thomas, and Vernon Thomas. 1990. "Just Let the Buffalo Roam." *Toronto Globe and Mail*, March 5, p. A7.

O'Gara, Geoffrey. 1989. "Beyond the Burn." *Sierra* 74 (January): 38–51.

Olsen, Russ. 1985. *Administrative History: Organizational Structure of the National Park Service—1917 to 1985.* Washington, D.C.: National Park Service.

Omang, Joannne. 1980. "Hill Clears Monumental Alaska Lands Bill." *Washington Post,* November 13, p. A5.

———. 1981. "Man with a Mission," *Washington Post,* March 9, p. A9.

Ovenden, Norm. 1990. "Logging to Continue: Company Lease Good for 12 Years." *Calgary Herald,* December 18, p. B5.

Parent, C. R., and F. E. Robeson. 1976. "An Economic Analysis of the River-Running Industry." Washington, D.C.: National Park Service.

Parks Canada. 1979. *Parks Canada Policy.* Parks Canada.

———. 1985. *Parks.* Parks Canada.

———. 1986. "National Park Warden Service." Parks Canada.

Passell, Peter. 1993. "Polls May Help Government Decide the Worth of Nature." *New York Times*, September 6, p. 1.

Patterson, Bruce. 1983. "Relief Not on the Horizon in National Parks." *Calgary Herald*, February 22, p. C7.

———. 1979. "Rents in National Parks May Soar." *Calgary Herald*, June 27, p. A1.

Peffer, E. Louise. 1951. *The Closing of the Public Domain.* Stanford University Press.

PEI Associates, Inc. 1989. "Development of Emission Inventories in and around National Park Service Units." Washington, D.C.: Department of the Interior.

Pfiffner, James P. 1987. "Political Appointees and Career Executives: The Democracy-Bureaucracy Nexus." *Public Administration Review* 47 (1): 57–65.

Philip, Tom. 1991. "The Sky's the Limit." *San Jose Mercury News*, November 11, p. 1A.

Philip, Tom, and Jeanne Huber. 1991. "Tent Cabin 'Ghettos' Get Booted in Yosemite Proposal." *San Jose Mercury News*, December 18, p. 2D.

Phillips, Don. 1992. "Hill Controversy Looms on Yosemite Concession." *St. Louis Post-Dispatch*, December 18, p. A10.

Pigou, Arthur C. 1920. *The Economics of Welfare*. 4th ed., 1962. London: MacMillan.

Pimlott, Douglas H. 1970. "Education and National Parks." In *Canadian Parks in Perspective*, edited by J. G. Nelson, 153–73. Quebec: Harvest House.

Platiel, Rudy. 1983. "Acid Rain Threatens Newest National Park." *Toronto Globe and Mail*, July 5, p. 8.

Poindexter, Joseph B. 1991. "The Fragile Balance." *Life* 14 (6): 88–90.

Postel, Sandra, and John C. Ryan. 1991. "Reforming Forestry." In *State of the World 1991*, edited by Lester R. Brown. Norton.

Pressman, Jeffrey L., and Aaron Wildavsky. 1984. *Implementation*. 3d ed. University of California Press.

Pritchard, Paul. 1991a. "The Best Idea America Ever Had." *National Geographic* 180 (2): 36–59.

———. 1991b. "Independence." *National Parks* 65 (November/December): 4.

———. 1992. "Perspectives." *National Parks* 66 (January/February): 4.

Pross, Paul. 1985. "Parliamentary Influence and the Diffusion of Power." *Canadian Journal of Political Science* 18 (2): 235–66.

Purser, Richard. 1982. "Proposal to Close National Parks Highway Is Dropped." *Calgary Herald,* October 12, p. D7.

Pyne, Stephen J. 1989. *Fire on the Rim*. New York: Ballantine Books.

Rauber, Paul. 1991. "Yosemite: Paradise Regained?" *Sierra* 76 (March/April): 24–28.

———. 1992. "The August Coup." *Sierra* 77 (January/February): 26–28.

Reich, Robert B. 1985. "Public Administration and Public Deliberation: An Interpretive Essay." *Yale Law Journal* 94 (June): 1617–41.

Reid, T. R. 1987. "Great Basin Park a First for Reagan." *Washington Post*, August 15, pp. A1, A11.

———. 1989. "Passive Policy on Natural Forest Fires Reaffirmed." *Washington Post*, June 2, p. A3.

Reisner, Marc P. 1986. *Cadillac Desert*. Viking.

Reuters News Service. 1966. "1972 Summer Olympics Go to Munich and Winter Games to Sapporo in Japan." *New York Times*, April 27, p. 54.

Ridenour, James. 1991. "Building on a Legacy." *National Parks* 65 (May/June): 22–23.

Riley, Michael. 1992. "Bighorn Gets over 900 Comments on Logging Plan." *Casper Star-Tribune*, August 29, p. B1.

Ripley, Randall B., and Grace A. Franklin. 1976. *Congress, the Bureaucracy, and Public Policy*. Homewood, Ill.: Dorsey Press.

Rosenbaum, Walter A. 1989. "The Bureaucracy and Environmental Policy." In *Environmental Politics and Policy*, edited by James P. Lester, 212–37. Duke University Press.

———. 1991. *Environmental Politics and Policy*. 2d ed. Washington, D.C.: CQ Press.

———. 1994. "The Clenched Fist and the Open Hand." In *Environmental Policy in the 1990s*, edited by Norman J. Vig and Michael E. Kraft, 2d ed., 121–43. Washington, D.C.: CQ Press.

Rourke, Francis E. 1984. *Bureaucracy, Politics and Public Policy*. 3d ed. Little, Brown and Company.

———. 1991. "Presidentializing the Bureaucracy: From Kennedy to Reagan." In *The Managerial Presidency*, edited by James P. Pfiffner, 123–34. Pacific Grove, Calif.: Brooks/Cole.

Rowe, Stan. 1990. *Home Place*. Edmonton: NeWest Press.

Rowley, William D. 1985. *U.S. Forest Service Grazing and Rangelands*. Texas A & M University Press.

Rowntree, R. A., and J. F. Orr. 1978. "The American Park Service." In *The Canadian National Parks: Today and Tomorrow Conference II*, edited by J. G. Nelson, 373–410. University of Waterloo.

Runte, Alfred. 1979. *National Parks: The American Experience*. University of Nebraska Press.

———. 1987. *National Parks: The American Experience*, 2d ed. University of Nebraska Press.

———. 1990. *Yosemite: The Embattled Wilderness*. University of Nebraska Press.

Russell, Dick. 1990. "Environmental Movers and Shakers." *American Forests* 96 (March/April): 24, 76.

Russell, Peter H. 1969. *The Supreme Court of Canada as a Bilingual and Bicultural Institution*. Ottawa: Queen's Printer.

Satchell, Michael. 1992. "Lonely Rangers in Paradise." *U.S. News & World Report*, January 13, pp. 29–30.

Savage, James D. 1991. "Saints and Cardinals in Appropriations Committees and the Fight against Distributive Politics." *Legislative Studies Quarterly* 16 (3): 329–47.

Sawatsky, John. 1991. *Mulroney: The Politics of Ambition*. Toronto: MacFarlane Walter & Ross.

Sax, Joseph L. 1980. *Mountains without Handrails: Reflections on the National Parks*. University of Michigan Press.

———. 1982. "The Compromise Called For." In *National Parks in Crisis*, edited by Eugenia Horstman Connally, 55–73. Washington, D.C.: National Parks and Conservation Association.

Sax, Joseph L., and Robert B. Keiter. 1988. "Glacier National Park and Its Neighbors: A Study of Federal Inter-Agency Cooperation." In *Our Common Lands: Defending the National Parks*, edited by David J. Simon, 175–241. Washington, D.C.: National Parks and Conservation Association.

Scace, R. C. 1968. *Banff: A Cultural Historical Study of Land Use and Management in a National Park Community to 1945*. University of Calgary: Studies in Land Use History, no. 2.

Schattschneider, E. E. 1960. *The Semisovereign People*. Holt, Rinehart and Winston.

Schmidt, William E. 1989. "From Yellowstone Ashes, New Life and Approach." *New York Times*, May 21, p. 26.

Schnoes, Roger, and Edward Starkey. 1978. *Bear Management in the National Park System*. Oregon State University.

Schullery, Paul. 1986. *The Bears of Yellowstone*. Boulder: Robert Rinehart, Inc. Publishers.

Schwartz, Mildred A. 1986. "Comparing United States and Canadian Public Policy: A Review of Strategies." *Policy Studies Journal* 14 (4): 566–79.

Scott, Myron L. 1991. "Defining NEPA Out of Existence." *Environmental Law Journal* 21: 807–38.

Seale, Ronald G. 1979. "It's Time to Realize National Parks Can't Be All Things to All People." *Calgary Herald*, July 3, p. A7.

Segal, Troy. 1991. "Rustic Luxury." *Business Week*, May 27, pp. 121–26.

Seligsohn-Bennett, Kyla. 1990. "Mismanaging Endangered and 'Exotic' Species in the National Parks." *Environmental Law Journal* 20: 415–40.

Sellars, Richard West. 1989. "Not Just Another Pretty Facade." *Washington Post*, April 9, p. B1.

———. 1992. "Tracks in the Wilderness." *Washington Post*, February 23, p. C5.

Selznick, Philip. 1949. *TVA and the Grass Roots*. Univerity of California Press.

———. 1957. *Leadership in Administration*. Evanston, Ill.: Row, Peterson & Co.

Shabecoff, Philip. 1988. "Report Finds Flaws in Fire Planning but Backs Policies as Basically Sound." *New York Times*, December 16, p. A22.

Shakespeare, William. 1988 ed. *Julius Caesar*. Cambridge University Press.

Shankland, Robert. 1970. *Steve Mather of the National Parks*. 3d ed. New York: Alfred A. Knopf.

Shanks, Bernard. 1984. *This Land Is Your Land*. San Francisco: Sierra Club Books.

Shelby, Bo, and Joyce M. Nielsen. 1976. "Motors and Oars in the Grand Canyon. River Contact Study Final Report." Washington, D.C.: National Park Service.

Shenan, Philip, and Philip Shabecoff. 1988. "New Civil War Battles." *New York Times*, November 2, p. 24.

Sholly, Dan R., and Steven M. Newman. 1991. *Guardians of Yellowstone*. New York: Morrow.

Shute, Nancy. 1992. "Leggo My Biosphere." *Outside* 17 (4): 26–28.

Simard, Carolle. 1988. "New Trends in Public Service Management." In *Canada under Mulroney*, edited by Andrew B. Gollner and Daniel Salee, 357–64. University of Toronto Press.

Simon, Dave. 1992. "Shortsighted on the Shenandoah." *Washington Post,* December 13, p. C8.

Skowronek, Stephen. 1982. *Building a New American State*. Cambridge University Press.

Smith, Adam. 1776. *The Wealth of Nations*, vol. 2. 1960 ed. London: Dent.

Smith, Zachary A. 1992. *The Environmental Policy Paradox*. Prentice-Hall.

Soden, Dennis L. 1991. "National Parks Literature of the 1980s: Varying Perspectives but Common Concerns." *Policy Studies Journal* 19 (3–4): 570–76.

Solow, Robert M. 1991. "Sustainability: An Economist's Perspective." In *Economics of the Environment*, edited by Robert Dorfman and Nancy Dorfman, 179–87, 3d ed. Norton.

Sonnichsen, Richard C. 1987. "Communicating Excellence in the FBI." In *Organizational Excellence*, edited by Joseph S. Wholey, 123–41. Lexington, Mass.: Lexington Books.

Sontag, Bill. 1990. *The First 75 Years*. Washington, D.C.: National Park Service.

Stanfield, Rochelle L. 1987. "Tilting on Development." *National Journal*, February 7, pp. 313–18.

Stegner, Wallace. 1989. "Our Common Domain." *Sierra* 74 (September/October): 42–47.

Stephenson, Marylee. 1983. *Canada's National Parks*. Prentice-Hall.

Stewart, Lillian, and Max Finkelstein. 1985. "Parks, National." *The Canadian Encyclopedia*, 1360–63. Edmonton: Hurtig Publishers.

Stockman, David A. 1986. *The Triumph of Politics*. Harper & Row.

Stroup, Richard L. 1990. "Rescuing Yellowstone from Politics: Expanding Parks while Reducing Conflict." In *The Yellowstone Primer*, edited by John A. Baden and Donald Leal, 169–84. San Francisco: Pacific Research Institute.

Struck, Doug. 1991. "Wolves: Plans to Return Wild Creatures to West Draw Howls from Many." *St. Louis Post-Dispatch*, July 21, p. 72G.

Struzik, Ed. 1990. "Timber Lease Adds to Fear about Future." *Calgary Herald*, December 29, p. A5.

Suro, Roberto. 1988. "At Yellowstone Park, Fires Echo in Debate on Rangers' Role." *New York Times*, June 4, p. B8.

Sutherland, Sharon L., and G. Bruce Doern. 1985. *Bureaucracy in Canada: Control and Reform*. University of Toronto Press.

Swem, Theodore R. 1970. " Planning of National Parks in the United States." In *Canadian Parks in Perspective*, edited by J. G. Nelson, 249–63. Quebec: Harvest House.

Taylor, Andrew. 1993. "President Will Not Use Budget to Rewrite Land-Use Laws." *CQ Weekly Report*, April 3, pp. 833–34.

Taylor, C. J. 1990. *Negotiating the Past*. McGill-Queen's University Press.

Taylor, Gregg W., and Doug Nixon. 1990. "An Unfinished Green Plan." *Maclean's*, June 4, p. 25.

Tellier, Paul M. 1990. "Public Service 2000: The Renewal of the Public Service." *Canadian Public Administration* 33 (2): 123–32.

Theberge, John B. 1976. "Ecological Planning in National Parks." In *Canada's Natural Environment*, edited by G. R. McBoyle and E. Sommerville, 194–216. Toronto: Methuen Publications.

———. 1981. "Guest Editorial." *Nature Canada* 10 (1): 23.

Thompson, D. N., A. J. Rogers, Jr., and F. Y. Borden. 1974. "Sound-Level Evaluations of Motor Noise." Washington, D.C.: National Park Service.

Tipple, Terence J., and J. Douglas Wellman. 1991. "Herbert Kaufman's Forest Ranger Thirty Years Later: From Simplicity and Homogeneity to Complexity and Diversity." *Public Administration Review* 51 (September/October): 421–28.

Trudeau, Pierre. 1972. *Conversation with Canadians*. University of Toronto Press.

Turque, Bill. 1991. "The War for the West." *Newsweek*, September 30, pp. 18–32.

Twomey, Steve. 1992. "Something Stinks about This Stable." *Washington Post*, December 14, p. D1.

Uhlenbrock, Tom. 1991. "Schober Favors Arch for VP Fair." *St. Louis Post-Dispatch*, July 14, p. 8D.

U.S. Congress. *Congressional Record*. 1916, 1940, 1946, 1987. Washington, D.C.

U.S. Congress, Conference Report on Appropriations. 1986. 99th Cong., 2d sess. *Making Appropriations for the Department of the Interior and Related Agencies.* Government Printing Office.

U.S. Congress, House Committee on Appropriations. 1983, 1984, 1985, 1986, 1987, 1988. *Department of the Interior and Related Agencies Appropriations Bill.* Government Printing Office.

U.S. Congress, House Committee on Interior and Insular Affairs. 1988. *Establishing a National Park System Review Board, and For Other Purposes.* 100th Cong., 2d sess. June 30. Rept. 100–742. Government Printing Office.

———. 1992. *National Parks and Public Lands Wilderness Management Act.* 102d Cong., 2d sess. Government Printing Office.

U.S. Congress, House Subcommittee on Environment, Energy, and Natural Resources Hearings. 1990. *Problems with Clean Air Act Protection for National Parks and Wilderness Areas.* 101st Cong., 2d sess. Government Printing Office.

U.S. Congress, Senate Committee on Energy and Natural Resources. 1981. Hearings on Amending the Land and Water Conservation Fund Act of 1965. May 7. 97th Cong., 1st sess. Government Printing Office.

U.S. Congress, Senate Committee on Energy and Natural Resources. 1982. Hearings. *Proposed Fiscal Year 1983 Budget Request.* 97th Cong., 2d sess. Government Printing Office.

U.S. Congress, Senate Subcommittee on Public Lands, National Parks, and Forests. 1986. *Entrance Fees and Resource Protection for Units of the National Park System.* 99th Cong., 2d sess. Government Printing Office.

———. 1987. *National Park Service Entrance Fees.* 100th Cong., 1st sess. Government Printing Office.

———. 1988. *Hearings on Current Fire Management Policies.* 100th Cong., 2d sess. Government Printing Office.

———. 1990. *Hearings on Concessions Policy of the National Park Service.* 101st Cong., 2d sess. Government Printing Office.

U.S. Department of the Interior. 1926. *Report of the Director of the National Park Service.* Government Printing Office.

———. 1990a. *Budget Justifications FY 1991.* Government Printing Office.

———. 1990b. *Report of the Task Force on National Park Concessions.* Department of the Interior.

———. 1992. *Information Packet.* Department of the Interior.

U.S. General Accounting Office. 1982. *Increasing Entrance Fees—National Park Service: Report to Congress.* GAO/CED-82-84. General Accounting Office.

———. 1987. *Parks and Recreation: Limited Progress Made in Documenting and Mitigating Threats to the Parks.* February. GAO/RCED-87-36. General Accounting Office.

———. 1991a. *Recreation Concessionaires Operating on Federal Lands.* GAO/T-RCED-91-16. General Accounting Office.

———. 1991b. *Status of Development at the Steamtown National Historic Site.* General Accounting Office.

———. 1992a. *Federal Lands: Oversight of Long-Term Concessioners.* March. GAO/RCED-92-128BR. General Accounting Office.

———. 1992b. *National Parks: Issues Involved in the Sale of the Yosemite National Park Concessioner*. September. GAO/RCED-92-232. General Accounting Office.

———. 1992c. *Policies and Practices for Determining Concessioners' Building Use Fees*. General Accounting Office.

U.S. National Park Service. 1972. *National Park System Plan*. National Park Service.

———. 1974. *Yellowstone Master Plan*. Department of the Interior.

———. 1976a. *Grand Canyon Final Master Plan*. Department of the Interior.

———. 1976b. *Guadalupe Mountains Proposed Master Plan*. Department of the Interior.

———. 1977a. *Grand Canyon Natural Resources Management Plan and Environmental Assessment*. Department of the Interior.

———. 1977b. *Second Annual Shenandoah Research Symposium*. Department of the Interior.

———. 1979a. *Colorado River Management Plan*. Department of the Interior.

———. 1979b. *Final Environmental Impact Statement of Proposed Colorado River Management Plan*. Department of the Interior.

———. 1980a. *Guadalupe Mountains Environmental Assessment*. Department of the Interior.

———. 1980b. *State of the Parks 1980*. National Park Service.

———. 1980c. *Statistical Report*. National Park Service.

———. 1980d. *Yosemite General Management Plan*. Department of the Interior.

———. 1983. *Shenandoah General Management Plan*. National Park Service.

———. 1986. *Twelve-Point Plan*. National Park Service.

———. 1987a. *National Park Service Career Management Concept*. Government Printing Office.

———. 1987b. "Yosemite Valley/El Portal Comprehensive Design." Department of the Interior.

———. 1988a. *Final Environmental Impact Statement and Development Concept Plan: Fishing Bridge Developed Area, Yellowstone National Park*. Department of the Interior.

———. 1988b. "FY 1987 Financial Report." National Park Service.

———. 1988c. *Management Policies*. National Park Service.

———. 1988d. *Natural Resources Assessment and Action Program Report*. National Park Service.

———. 1989a. *Colorado River Management Plan*. Department of the Interior..

———. 1989b. *National Parks Index 1989*. Government Printing Office..

———. 1990a. *Endangered Species in the National Parks*. National Park Service.

———. 1990b. *State of the Rocky Mountain Region*. Appendix A. National Park Service.

———. 1990c. "Western Regional Office Organizational Chart." National Park Service.

———. 1991a. *Criteria for Parklands*. National Park Service.

———. 1991b. *Goals and Objectives of the Rocky Mountain Regional Office*. Department of the Interior.

———. 1991c. *Parks Etc*. National Park Service.

———. 1991d. *Yellowstone Statement for Management*. Department of the Interior.

————. 1992a. "Air Quality Fact Sheet." Shenandoah National Park: National Park Service.

————. 1992b. *National Parks for the 21st Century: The Vail Agenda.* National Park Service.

————. 1992c. "Steamtown NHS." National Park Service.

————. 1992d. *Yellowstone National Park Wildland Fire Management Plan.* Department of the Interior.

————. Various years. *Statistical Abstract.* National Park Service.

Val, Erik. 1987. "Socioeconomic Impact Assessment, Regional Integration, Public Participation, and New National Park Planning in Canada." In *Social Science in Natural Resource Management Systems,* edited by M. L. Miller, R. P. Gale, and P. J. Brown, 129–48. Boulder: Westview Press.

Van der Wert, Martin, and Jeff Barber. 1992. "Canyon Flow Limits Signed into Law by Bush in Likely Campaign Move." *Arizona Republic,* October 31, p. A2.

Van Wagner, C. E., and I. R. Methuen. 1980. *Fire in the Management of Canada's National Parks: Philosophy and Strategy.* Chalk River: Canadian Forestry Service.

Vig, Norman J. 1990. "Presidential Leadership: From the Reagan to the Bush Administration." In *Environmental Policy in the 1990s,* edited by Norman J. Vig and Michael E. Kraft, 33–58. Washington, D.C.: CQ Press.

Vogel, David. 1993. "Representing Diffuse Interests in Environmental Policymaking." In *Do Institutions Matter?* edited by R. Kent Weaver and Bert A. Rockman, 237–71. Brookings.

Voight, William, Jr. 1976. *Public Grazing Lands.* Rutgers University Press.

Wade, Betsy. 1991. "Dial 1–800: Campers Can Reserve Park Sites Using Ticketron Number." *St. Louis Post-Dispatch,* March 24, p. 2T.

————. 1993. "National Parks Are Battling the Budget Crunch." *St. Louis Post-Dispatch,* May 9, p. 4T.

Waiser, William. 1989. *Saskatchewan's Playground.* Saskatoon: Fifth House.

Walker, William. 1991. "National Park Plan in Peril, Groups Say." *Toronto Star,* October 9, p. A4.

————. 1992. "Feds Falling behind on National Parks Program." *Vancouver Sun,* January 18, p. A13.

Warwick, Donald P. 1975. *A Theory of Public Bureaucracy.* Harvard University Press.

Washington Post. 1988. "Burn Policy in National Parks 'Sound'." *St. Louis Post-Dispatch,* December 18, p. 12.

Watts, Ronald L. 1991. "Canadian Federalism in the 1990s—Once More in Question." *Publius* 21 (3): 169–90.

Weaver, R. Kent. 1992. "Political Institutions and Canada's Constitutional Crisis." In *The Collapse of Canada?* edited by R. Kent Weaver, 7–75. Brookings.

Weaver, R. Kent, and Bert A. Rockman. 1993. "Assessing the Effects of Institutions." In *Do Institutions Matter?* edited by R. Kent Weaver and Bert A. Rockman, 1–41. Brookings.

Werier, Val. 1991. "Riding Mountain's Remarkable Bears Face Threat." *Winnipeg Free Press,* May 25, p. 6.

Whitaker, John C. 1976. *Striking a Balance*. Washington, D.C.: American Enterprise Institute.

Wholey, Joseph S. 1987. "Toward Excellence in Government." In *Organizational Excellence*, edited by Joseph S. Wholey, 179–84. Lexington, Mass.: Lexington Books.

Wilson, James Q. 1980. "The Politics of Regulation." In *The Politics of Regulation*, edited by James Q. Wilson, 357–94. Basic Books.

———. 1989. *Bureaucracy*. Basic Books.

Wilson, Seymour, and O. P. Dwivedi. 1982. "Introduction." In *The Administrative State in Canada*, edited by O. P. Dwivedi, 3–15. University of Toronto Press.

Winks, Robin W. 1993. "National Parks Aren't Disneylands." *New York Times*, April 19, p. A11.

Winning, Bruce. 1982. "Banff and Jasper National Parks: Poachers Deplete Stock." *Calgary Herald*, July 21, p. C1.

Wirth, Conrad L. 1980. *Parks, Politics, and the People*. University of Oklahoma Press.

Wood, B. Dan, and Richard W. Waterman. 1991. "The Dynamics of Political Control of the Bureaucracy." *American Political Science Review* 85 (September): 801–28.

Worsnop, Richard L. 1993a. "National Parks: Should Parks Limit Visitors, or Try to Meet Demands?" *CQ Researcher*, May 28, pp. 457–79.

———. 1993b. "Saving Our Parks." *Knoxville News-Sentinel*, June 13, p. F1.

Wren, Christopher S. 1985. "Canada Announces Plan to Reduce Its Acid Rain." *New York Times*, March 7, p. A6.

Wuerthner, George. 1989. "Burning Issues." *Sierra* 74 (January): 43.

Ybarra, Michael J. 1990. "Concessions Policy for Parks Revised." *Washington Post*, July 20, p. A6.

Zuckerman, Seth. 1992. "New Forestry? New Hype . . . " *Sierra* 77 (February): 40–45.

Index

269